METHODS
IN
YEAST GENETICS

A Cold Spring Harbor Laboratory Course Manual

2005 Edition

METHODS
IN
YEAST GENETICS

A Cold Spring Harbor Laboratory Course Manual

2005 Edition

David C. Amberg
SUNY Upstate Medical University

Daniel J. Burke
University of Virginia Medical Center

Jeffrey N. Strathern
National Cancer Institute

CSHL PRESS
COLD SPRING HARBOR LABORATORY PRESS

METHODS IN YEAST GENETICS

A Cold Spring Harbor Laboratory Course Manual
2005 Edition

Printed in the United States of America

Publisher	John Inglis
Acquisition Editor	David Crotty
Project Coordinator	Maryliz Dickerson
Permissions Coordinator	Maria Fairchild
Production Editor	Rena Steuer
Desktop Editor	Susan Schaefer
Production Manager	Denise Weiss
Cover Designer	Ed Atkeson

Front cover: The cell cycle of yeast. DNA is shown in magenta, microtubules in green, the spindle pole body is dark blue, and the cell outline is blue. Photo credit: D.W. Hailey, T.N. Davis, and E.G.D. Muller, Yeast Resource Center and Department of Biochemistry, University of Washington. Reprinted from *Methods in Cell Biology*, vol. 67, Centrosomes and Spindle Pole Bodies (R.E. Palazzo and T.N. Davis, eds.), cover photo, Copyright 2001 with permission from Elsevier.

Back cover: Haploid (*top*) and diploid (*bottom*) yeast cells were stained to visualize the actin cytoskeleton with rhodamine-phalloidin (blue) and the bud scars with calcofluor (magenta). The stained cells were imaged by D. Amberg on a Deltavision deconvolution microscope. Optical sections were captured in a Z-series at 0.2-μm intervals, out-of-focus light was removed by reiterative deconvolution, and three-dimensional projections were calculated. The images shown are stereo pairs of the projections rotated 5° in the y axis. Image courtesy of D.C. Amberg, SUNY Upstate Medical University.

Library of Congress Cataloging-in-Publication Data

Amberg, David C.
Methods in yeast genetics : a Cold Spring Harbor Laboratory course manual / David C. Amberg, Daniel J. Burke, Jeffrey Strathern.-- 2005 ed.
p. ; cm.
Rev. ed. of: Methods in yeast genetics / Dan Burke, Dean Dawson, Tim Stearns, 2000 ed. 2000.
Includes bibliographical references and index.
ISBN 0-87969-728-8 (pbk. : alk. paper)

[DNLM: 1. Yeasts--genetics--Laboratory Manuals. QW 25 A491m 2005] I. Burke, Dan, 1954- II. Strathern, Jeffrey N. III. Burke, Dan, 1954- Methods in yeast genetics. IV. Cold Spring Harbor Laboratory. V. Title.

2004028765

10 9 8 7 6 5 4 3 2 1

Students and researchers using the procedures in this manual do so at their own risk. Cold Spring Harbor Laboratory makes no representations or warranties with respect to the material set forth in this manual and has no liability in connection with the use of these materials. All registered trademarks, trade names, and brand names mentioned in this book are the property of the respective owners. Readers should please consult individual manufacturers and other resources for current and specific product information.

With the exception of those suppliers listed in the text with their addresses, all suppliers mentioned in this manual can be found on the BioSupplyNet Web site http://www.biosupplynet.com

All World Wide Web addresses are accurate to the best of our knowledge at the time of printing.

Procedures for the humane treatment of animals must be observed at all times. Check with the local animal facility for guidelines.

Certain experimental procedures in this manual may be the subject of national or local legislation or agency restrictions. Users of this manual are responsible for obtaining the relevant permissions, certificates, or licenses in these cases. Neither the authors of this manual nor Cold Spring Harbor Laboratory assume any responsibility for failure of a user to do so.

The polymerase chain reaction process is covered by certain patent and proprietary rights. Users of this manual are responsible for obtaining any licenses necessary to practice PCR or to commercialize the results of such use. COLD SPRING HARBOR LABORATORY MAKES NO REPRESENTATION THAT USE OF THE INFORMATION IN THIS MANUAL WILL NOT INFRINGE ANY PATENT OR OTHER PROPRIETARY RIGHT.

All Cold Spring Harbor Laboratory Press publications may be ordered directly from Cold Spring Harbor Laboratory Press, 500 Sunnyside Blvd., Woodbury, New York 11797-2924. Phone: 1-800-843-4388 in Continental U.S. and Canada. All other locations: (516) 422-4100. FAX: (516) 422-4097. E-mail: cshpress@cshl.edu. For a complete catalog of all Cold Spring Harbor Laboratory Press publications, visit our World Wide Web site http://www.cshlpress.com./

Contents

Appendices

Preface

This laboratory course manual incorporates significant portions of the manuals used in previous Cold Spring Harbor Laboratory yeast genetics courses. Although most of the experiments have now been revised and several new techniques have been added, the basic structure of this course is the same as it has been for the past 30 years. We are indebted to our predecessors, Fred Sherman, Gerry Fink, Jim Hicks, Cal McLaughlin, Brian Cox, Mark Rose, Fred Winston, Phil Hieter, Susan Michaelis, Aaron Mitchell, Alison Adams, Chris Kaiser, Dan Gottschling, Tim Stearns, Dean Dawson, and Orna Cohen-Fix for their teaching and for making this course an important part of the yeast community. We also thank Mike Cherry for his invaluable assistance with the genetic and physical maps.

Dave Amberg
Dan Burke
Jeff Strathern

Introduction

Since the last edition of this manual was published, *Saccharomyces cerevisiae* has continued to be utilized as the testing ground for new genomics and proteomics technologies. Although it can be difficult to incorporate some of these methods into a three-week laboratory course, some of the most powerful methods have been included in the Experiments section or detailed protocols have been included in the Techniques and Protocols section where possible. However, many of the latest technologies are only a starting point that necessitate subsequent, classical yeast genetic analyses. For this reason, much remains fundamentally unchanged from previous editions, recognizing that the power of this model system lies in its sexual cycle, the ability to do tetrad analysis, and a high rate of homologous recombination. We hope that this manual will continue to serve as a resource for properly performing yeast genetic experiments.

Some of the additions to the Experiments section include more yeast vital stains such as FM4-64 staining of endocytic compartments and the visualization of green fluorescent protein (GFP)-tagged proteins in Experiment 1. In Experiment VII (Gene Replacement) we have adopted synthetic lethal analysis by systematic genetic analysis (SGA), as first described by the Boone lab (Tong et al. 2001). The Gene Replacement experiment has been modernized to include new drug-resistance markers, such as kanMX6, as well as markers from related yeasts less likely to recombine with the *S. cerevisiae* locus, such as *S. pombe* $his5^{+}$. High-copy suppression has been added to the *ras2* suppressors experiment (VIII) recognizing the method's contributions to understanding signaling pathways. The two-hybrid experiment has been replaced with another two-hybrid based method that identifies separation of function mutations (Experiment XI). In the Techniques and Protocols section, we have added a Tandem Affinity Protein tag protein purification protocol (Techniques and Protocols #5) adapted from the procedure first developed in the Seraphin lab (Rigaut et al. 1999), and a technique for gene disruption by double-fusion polymerase chain reaction (PCR) (Techniques and Protocols #14). We have included a section on handling large collections of yeast strains generated in systematic studies and we use the yeast deletion collection as an example. We have included simpler and more reliable protocols for extracting proteins from yeast and for colony PCR. We also added an improved flow cytometry protocol using the dye SYTOX Green as described by Steve Haase (Haase and Reed 2002) and some basic *E. coli* protocols required for the new or altered experiments (DNA miniprep and transformation). Finally, given the explosion of traditional and Web-based tools, we have greatly expanded the list of useful books and Web sites found at the end of this Introduction.

Genetic investigations of yeast were initiated by Winge and his co-workers in the mid-1930s. Approximately ten years later, Lindegren and his colleagues also began extensive studies. These two groups are responsible for uncovering the general principles and much of the basic methodology of yeast genetics. Today, yeast is widely recognized as an ideal eukaryotic microorganism for biochemical and genetic studies. Although yeasts have greater genetic complexity than bacteria, they still share many of the technical advantages that permitted rapid progress in the molecular genetics of prokaryotes and their viruses. Some of the properties that make yeast particularly suitable for genetic studies include the existence of both stable haploid and diploid cells, rapid growth, clonability, a particularly high rate of homologous recombination in conjunction with a low rate of nonhomologous end joining, the ease of replica-plating, mutant isolation, and the ability to isolate each haploid product of meiosis by microdissection of a tetrad ascus. Yeast has been successfully employed for the study of all areas of genetics, such as mutagenesis, recombination, chromosome segregation, and gene action and regulation, as well as aspects distinct to eukaryotic systems, such as mitochondrial genetics.

DNA transformation has made yeast particularly accessible to gene cloning and genetic engineering techniques. Structural genes corresponding to virtually any genetic trait can be identified by complementation from plasmid libraries. DNA is introduced into yeast cells either as replicating molecules or by integration into the genome. In contrast to most other organisms, integrative recombination of transforming DNA in yeast proceeds primarily via homologous recombination. This permits efficient targeted integration of DNA sequences into the genome. Homologous recombination coupled with high levels of gene conversion has led to the development of techniques for the direct replacement of normal chromosomal loci with genetically engineered DNA sequences. This ease of performing direct gene replacement is unique among eukaryotic organisms and has been extensively exploited in every aspect of yeast genetics, cell biology, physiology, and biochemistry. Many of these modern genetic techniques are reviewed in volume 194 of *Methods in Enzymology* (Guthrie and Fink 1991), volumes 350 and 351 of *Methods in Enzymology* (Guthrie and Fink 2002), and by Rose (1995).

Recent advances in modern yeast genetics have come from determining the complete DNA sequence of the *Saccharomyces cerevisiae* genome and from the construction of a complete collection of deletion mutants. The assembly of this sequence into a public sequence database has made yeast one of the premier organisms for detailed analysis of eukaryotic cellular function and genome organization. Several Web sites provide convenient and organized access to the sequence database. Two of the more popular sites are the *Saccharomyces* Genome Database (SGD) and the Munich Information Center for Protein Sequences (MIPS). Each of these sources augments the sequence database by providing powerful search functions for analyzing the annotated genome, as well as helpful links to relational databases, such as literature cross references or the protein structure of yeast gene products. In addition, the Yeast Protein Database (YPD) includes each yeast protein and predicted open reading frame compiled with information about its structure and the phenotypes associated with mutations in its gene.

One of the most useful collections of information on *S. cerevisiae* biology is found in two companion series of reviews entitled *The Molecular Biology of the Yeast* Saccharomyces (Strathern et al. 1981, 1982) and *The Molecular and Cellular Biology of the Yeast* Saccharomyces (Broach et al. 1991; Jones et al. 1992; Pringle et al. 1997). Although a few chapters in the later series update progress made since the first series was published, each volume has detailed reviews unavailable anywhere else that synthesize the vast literature of yeast biology. All five of these books are essential for the library of a yeast biologist. In addition, *The Early Days of Yeast Genetics* (Hall and Linder 1993) provides a historical perspective on why yeast became a model organism and traces the development of the yeast community's influence on modern genetics and eukaryotic molecular biology.

This course focuses exclusively on the baker's yeast, *S. cerevisiae*. After completing the course, you should be able to carry out all of the techniques commonly employed by yeast geneticists and follow the literature with greater ease. Except for the dissection of asci, most of the methods do not differ significantly from the methods used with other microorganisms, and the skills should be rapidly acquired with little practice. Experiments will be conducted in pairs, and whenever possible, an investigator more familiar with microbiological techniques will be assigned to a less-experienced partner. Since most of the experiments will be initiated at the outset of the course, it is advisable to read the entire manual thoroughly. Please note that some experiments span many days because of extended periods of incubation.

We wish to emphasize that some of the procedures in this manual have been condensed to save time and are not necessarily standard for research purposes. For example, mutants are usually purified by subcloning the initial isolates. Also, some of the techniques may not be directly applicable to your research problems, but they have been included to illustrate general principles and methods.

HIGHLY RECOMMENDED READING

Bartel P.L. and Fields S. 1997. *The yeast two-hybrid system*. Oxford University Press, United Kingdom.

Broach J.R., Jones E.W., and Pringle J.R. 1991. *The molecular and cellular biology of the yeast* Saccharomyces. I. *Genome dynamics, protein synthesis, and energetics*. Cold Spring Harbor Laboratory Press, Cold Spring Harbor, New York.

Campbell I. and Duffus J.H. 1988. *Yeast: A practical approach*. IRL Press, Oxford.

Fincham J.R.S., Day P.R., and Radford A. 1979. *Fungal genetics*. University of California Press, Berkeley.

Guthrie C. and Fink G.R., eds. 1991. Guide to yeast genetics and molecular biology. *Methods Enzymol.*, vol. 194.

———. 2002. Guide to yeast genetics and molecular and cell biology. *Methods Enzymol.*, vol. 350, Part B.

———. 2002. Guide to yeast genetics and molecular and cell biology. *Methods Enzymol.*, vol. 351, Part C.

Haase S.B. and Reed S.I. 2002. Improved flow cytometric analysis of the budding yeast cell cycle. *Cell Cycle* **1:** 132–136.

Hall M.N. and Linder P. 1993. *The early days of yeast genetics.* Cold Spring Harbor Laboratory Press, Cold Spring Harbor, New York.

Johnston J.R. 1994. *Molecular genetics of yeast: A practical approach.* IRL Press, Oxford.

Jones E.W., Pringle J.R., and Broach J.R. 1992. *The molecular and cellular biology of the yeast* Saccharomyces. II. *Gene expression.* Cold Spring Harbor Laboratory Press, Cold Spring Harbor, New York.

Pringle J.R., Broach J.R., and Jones E.W. 1997. *The molecular and cellular biology of the yeast* Saccharomyces. III. *Cell cycle and cell biology.* Cold Spring Harbor Laboratory Press, Cold Spring Harbor, New York.

Rigaut G., Shevchenko A., Rutz B., Wilm M., Mann M., and Seraphin B. 1999. A generic protein purification method for protein complex characterization and proteome exploration. *Nat. Biotechnol.* **10:** 1030–1032.

Rose M.D. 1995. Modern and post-modern genetics in Saccharomyces cerevisiae. In *The yeasts,* 2nd edition (ed. A.E. Wheals et al.), vol. 6, pp. 69–120. Academic Press, New York.

Strathern J.N., Jones E.W., and Broach J.R., eds. 1981. *The molecular biology of the yeast* Saccharomyces: *Life cycle and inheritance.* Cold Spring Harbor Laboratory, Cold Spring Harbor, New York.

———. 1982. *The molecular biology of the yeast* Saccharomyces: *Metabolism and gene expression.* Cold Spring Harbor Laboratory, Cold Spring Harbor, New York.

Tong A.H., Evangelista M., Parsons A.B., Xu H., Bader G.D., Page N., Robinson M., Raghibizadeh S., Hogue C.W., Bussey H., Andrews B., Tyers M., and Boone C. 2001. Systematic genetic analysis with ordered arrays of yeast deletion mutants. *Science* **294:** 2364–2368.

RECOMMENDED YEAST WEB SITES

The Brown Lab Microarray Resource
http://cmgm.stanford.edu/pbrown/

The Definitive Yeast Transformation Homepage
http://www.umanitoba.ca/faculties/medicine/biochem/gietz/Trafo.html

European *Saccharomyces cerevisiae* Archive for Functional Analysis
http://www.uni-frankfurt.de/fb15/mikro/euroscarf/col_index.html

GRID (database of Genetic and Physican Interactions and Osprey Network Visualization Software)
http://biodata.mshri.on.ca/grid/servlet/Index

Incyte Proteome BioKnowledge Library
http://www.incyte.com/control/researchproducts/insilico/proteome

Munich Information Center for Protein Sequences
http://mips.gsf.de/genre/proj/yeast/index.jsp

Saccharomyces Genome Database
http://www.pathway.yeastgenome.org/

TRIPLES (database of Transposon-Insertion Phenotypes, Localization, and Expression in *Saccharomyces*)
http://ygac.med.yale.edu/triples/

The University of Washington Yeast Resource Center
http://depts.washington.edu/~yeastrc/

Yeast GFP Fusion Localization Database
http://yeastgfp.ucsf.edu/

Genetic Nomenclature

CHROMOSOMAL GENES

Early recommendations for the nomenclature and conventions used in yeast genetics have been summarized by Sherman and Lawrence (1974) and Sherman (1981). Whenever possible, gene symbols are consistent with the proposals of Demerec et al. (1966) and are designated by three italicized letters (e.g., *arg*). Contrary to the proposals of Demerec et al. (1966), the genetic locus is identified by a number (not a letter) following the gene symbol (e.g., *arg2*). Dominant alleles are denoted by using uppercase italics for all three letters of the gene symbol (e.g., *ARG2*). Lowercase letters symbolize the recessive allele (e.g., the auxotroph *arg2*). Wild-type genes are designated with a superscripted plus sign (*sup6*$^+$ or *ARG2*$^+$). Alleles are designated with a number separated from the locus number by a hyphen (e.g., *arg2-14*). Locus numbers are consistent with the original assignments; however, allele numbers may be specific to a particular laboratory.

Phenotypic designations are sometimes denoted by cognate symbols in Roman type followed by a superscripted plus or minus sign. For example, the independence from and requirement for arginine can be denoted by Arg$^+$ and Arg$^-$, respectively.

The following examples illustrate the conventions used in the genetic nomenclature for *S. cerevisiae:*

ARG2	A locus or dominant allele
arg2	A locus or recessive allele that produces a requirement for arginine as the phenotype
ARG2$^+$	The wild-type allele of this gene
arg2-9	A specific allele or mutation at the *ARG2* locus
Arg$^+$	A strain that does not require arginine
Arg$^-$	A strain that requires arginine
Arg2p	Designation for the protein product of the *ARG2* gene

There are a number of exceptions to these general rules. Gene clusters, complementation groups within a gene, or domains within a gene that have different properties can be designated by uppercase letters following the locus number (e.g., *his4A, his4B*).

The extensive use of recombinant DNA techniques in yeast has introduced a nomenclature that pertains to gene insertions, gene fusions, and plasmids:

ARG2::LEU2	An insertion of the *LEU2* gene at the *ARG2* locus where the insertion does not disrupt *ARG2* function
arg2::LEU2	An insertion of the *LEU2* gene at the *ARG2* locus where the insertion disrupts *ARG2* function
arg2-101::LEU2	An insertion of the *LEU2* gene at the *ARG2* locus where the insertion disrupts *ARG2* function and the disruption allele is specified
arg2Δ0::LEU2	An insertion of the *LEU2* gene at the *ARG2* locus where the insertion disrupts *ARG2* function by an exact replacement of the entire *ARG2* open reading frame
cyc1-arg2	A gene fusion between the *CYC1* gene and *ARG2* where neither gene is functional
P_{cyc1}-*ARG2*	A gene fusion between the *CYC1* gene promoter and *ARG2* where the *ARG2* gene is functional
[YCp-*ARG2*]	A centromere plasmid carrying a functional *ARG2* locus
[pCK101]	Designation for a specific plasmid whose structure is given elsewhere

Although superscripts should be avoided, it is sometimes expedient to distinguish genes conferring resistance or sensitivity by a superscripted R or S, respectively. For example, the genes controlling resistance to canavanine sulfate (*CAN1*), copper sulfate (*CUP1*), and their sensitive alleles can be denoted, respectively, as $can^{R}1$, $CUP^{R}1$, $CAN^{S}1$, and $cup^{S}1$.

Wild-type and mutant alleles of the mating type and related loci do not follow the standard rules. The two wild-type alleles of the mating-type locus are designated *MAT***a** and *MATα*. Note that the lowercase "**a**" is bold and not italicized and the "*α*" is italicized but not bold. The two complementation groups of the *MATα* locus are denoted *MATα1* and *MATα2*. Mutations of the *MAT* genes are denoted, e.g., *mat***a***-1* and *matα1-1*. The wild-type homothallic alleles at the *HMR* and *HML* loci are denoted *HMR***a**, *HMRα*, *HML***a**, and *HMLα*. Mutations at these loci are denoted, e.g., *hmr***a***-1* and *hmlα-1*. The mating phenotypes of *MAT***a** and *MATα* cells are denoted simply **a** and *α*, respectively.

Dominant and recessive suppressors should be denoted, respectively, by three uppercase or lowercase letters followed by a locus designation (e.g., *SUP4, SUF1, sup35, suf11*). In some instances, UAA suppressors and UAG suppressors are further designated o and a, respectively, following the locus. For example, *SUP4*-o refers to suppressors of the *SUP4* locus that insert tyrosine residues at UAA sites and *SUP4*-a refers to suppressors of the same *SUP4* locus that insert tyrosine residues at UAG sites. The corresponding wild-type locus coding for the normal tyrosine tRNA and lacking suppressor activity can be referred to as $sup4^{+}$. Thus, the nomenclature describing suppressor and wild-type alleles in yeast is unrelated to the bacterial nomenclature. For example, an ochre *E. coli* suppressor that inserts tyrosine residues at both UAA and UAG sites is denoted as su_4^+, and the wild-type locus coding for the normal tyrosine tRNA and lacking suppressor activity can be referred to as Su_4, su_4^-, or *supC*.

For most structural genes that code for proteins, the functional wild-type allele is usually dominant to the mutant form of a gene. In yeast, the convention for dominant genes utilizes italic symbols such as *HIS4* and *LEU2*. Because the sites of recessive mutations are usually used for genetic mapping, published chromosome maps usually contain the mutant form of the gene. For example, chromosome III contains *his4* and *leu2*, whereas chromosome IX contains *SUP22* and *FLD1*. Because capital letters are used to represent dominant wild-type genes that control the same character (e.g., *SUC1, SUC2*), and because the dominant forms are used in genetic mapping, such chromosomal loci are denoted with uppercase letters on genetic maps. In addition, uppercase letters are used to designate certain DNA segments whose locations have been determined by a combination of recombinant DNA techniques and classical mapping procedures (e.g., *RDN1*, the segment encoding ribosomal RNA).

NON-MENDELIAN DETERMINANTS

Where necessary, non-Mendelian genotypes can be distinguished from chromosomal genotypes by enclosure in brackets. Whenever applicable, it is advisable to use the above rules for designating non-Mendelian genes and to avoid the use of Greek letters. However, when referring to an entire non-Mendelian element, it is best to either retain the original symbols [ρ^+], [ρ^-], [ψ^+], and [ψ^-] or use their transliteration, [*rho*$^+$], [*rho*$^-$], [*PSI*$^+$], and [*psi*$^-$], respectively. Detailed designations for mitochondrial mutants have been presented by Dujon (1981) and Grivell (1984, 1990) and for killer strains by Wickner (1981). Unlike the other non-Mendelian determinants, [*PSI*$^+$] and [*URE3*] are not based on different states of a nucleic acid; rather, the [*PSI*$^+$] and [*URE3*] traits result from heritable conformational states of proteins. The unusual behavior of these traits can be explained by the prion hypothesis, which has been used to explain infectious neurodegenerative diseases such as scrapie in mammals (Lindquist 1997). [*PSI*$^+$] corresponds to a heritable conformational state of the translation termination factor Sup35p, and [*URE3*] corresponds to a heritable inactive state of *URE2*, a gene whose product is involved in nitrogen regulation. The known non-Mendelian determinants in yeast are listed in Table 1.

Table 1. Non-Mendelian determinants of yeast

Wild type	Mutant variant	Element	Mutant trait
[ρ^+]	[ρ^-]	Mitochondrial DNA	Respiration deficiency
[*KIL*-k$_1$]	[*KIL*-o]	RNA plasmid	Sensitive to killer toxin
[*cir*$^+$]	[*cir*^{o}]	2μ plasmid	None
[*psi*$^-$]	[*PSI*$^+$]	Prion form of Sup35p	Enhanced suppression of nonsense codons
[*ure3*$^-$]	[*URE3*]	Prion form of Ure2p	Unregulated ureidosuccinate uptake

GENETIC BACKGROUNDS

The genetic background from which a *S. cerevisiae* strain is derived is an often hidden aspect of the genotype that should be taken into account when designing experiments. Most strains used in modern genetic studies come from one of a small set of genetic backgrounds, including S288C, X2180, A364a, W303a, Σ1278b, AB972, SK1, and FL100. The genealogies of some of these backgrounds have recently been reconstructed from records of crosses that were carried out in the 1940s between wild yeasts and brewing strains (Mortimer and Johnston 1986). This analysis shows that although most backgrounds share a common ancestry, a significant degree of genetic heterogeneity has been introduced by outcrossing. In practice, crosses between distantly related strains often give inviable combinations of alleles leading to many inviable spores, whereas crosses between strains from the same background usually give >95% viable spores. The S288C and A364a genetic backgrounds have similar genealogies and in crosses give a high frequency of spore viability, but an analysis of genomic sequences from these strains reveals an average of 3.4 nucleotide sequence differences per kilobase of genomic DNA. Thus, even apparently closely related strains can differ at a very large number of sites.

Allelic differences between strain backgrounds can seriously influence the outcome of many different kinds of experiments, and it is best to avoid genetic heterogeneity as much as possible by using a single genetic background. At the beginning of a new mutant hunt, it is worth considering which strain background to use—usually the background used by most investigators in the same field is the best choice. S288C is probably the most commonly used background; however, other backgrounds offer distinct advantages for particular types of experiments. For example, Σ1278b will form pseudohyphae, whereas S288C will not, and SK1 sporulates much more rapidly than S288C. It is often necessary to move a desired mutation from one background into another. Ideally, this can be done using recombinant plasmids and the methods for gene replacement described in Experiment VII. Mutations that have not been cloned can be moved by backcrossing to the desired strain background (usually, successive backcrosses are carried out until a clear 2:2 pattern of segregation for the desired trait has been achieved).

REFERENCES

Demerec M., Adelberg E.A., Clark A.J., and Hartman P.E. 1966. A proposal for a uniform nomenclature in bacterial genetics. *Genetics* **54:** 61–76.

Dujon B. 1981. Mitochondrial genetics and functions. In *The molecular biology of the yeast* Saccharomyces: *Life cycle and inheritance* (ed. J.N. Strathern et al.), pp. 505–635. Cold Spring Harbor Laboratory, Cold Spring Harbor, New York.

Grivell L.A. 1984. Restriction and genetic maps of yeast mitochondrial DNA. In *Genetic maps*, 3rd edition (ed. S.J. O'Brien), pp. 234–247. Cold Spring Harbor Laboratory, Cold Spring Harbor, New York.

———. 1990. Mitochondrial DNA in the yeast *Saccharomyces cerevisiae*. In *Genetic maps*, 5th edition (ed. S.J. O'Brien), pp. 3.50–3.57. Cold Spring Harbor Laboratory Press, Cold Spring Harbor, New York.

Lindquist S. 1997. Mad cows meet psi-chotic yeast: The expansion of the prion hypothesis. *Cell* **89:** 495–498.

Mortimer R.K. and Johnston J.R. 1986. Genealogy of principal strains of the yeast genetic stock center. *Genetics* **113:** 35–43.

Sherman F. 1981. Genetic nomenclature. In *The molecular biology of the yeast* Saccharomyces: *Life cycle and inheritance* (ed. J.N. Strathern et al.), pp. 639–640. Cold Spring Harbor Laboratory, Cold Spring Harbor, New York.

Sherman F. and Lawrence C.W. 1974. *Saccharomyces*. In *Handbook of genetics: Bacteria, bacteriophages, and fungi* (ed. R.C. King), vol. 1, pp. 359–393. Plenum Press, New York.

Wickner R.B. 1981. Killer systems in *Saccharomyces cerevisiae*. In *The molecular biology of the yeast* Saccharomyces: *Life cycle and inheritance* (ed. J.N. Strathern et al.), pp. 415–444. Cold Spring Harbor Laboratory, Cold Spring Harbor, New York.

Looking at Yeast Cells

Yeast cells are approximately 5 µm in diameter, and many of their important features can be seen in the light microscope. It is good laboratory practice to routinely examine cultures under phase microscopy for indications of the physiological state of the cells and for evidence of contamination. Much of modern yeast cell biological work involves more sophisticated examination of yeast cells stained with protein-specific antibodies, fluorescent dyes that specifically associate with certain organelles, or more recently, the use of fusions of green fluorescent protein (GFP) to a yeast protein of interest. This experiment provides examples of the standard types of light microscopy that are used in the examination of yeast cells.

EXAMINATION OF GROWING CULTURES

Growth Properties

Saccharomyces cerevisiae cells grow by budding. A cell that gives rise to a bud is called a mother cell, and the bud is sometimes referred to as the daughter cell. A new bud emerges from a mother cell close to the beginning of the cell cycle and continues to grow until it separates from the mother cell at the end of the cell cycle. Because all of the growth of a yeast cell is concentrated in the bud, and because this growth is essentially continuous, the size of the bud gives an approximate indication of the position of a given cell in the cell cycle. An exponentially growing culture of yeast cells has approximately one-third unbudded cells, one-third cells with a small bud, and one-third cells with a large bud. When cells in a growing culture use up the available nutrients, they stop growing by arresting in the cell cycle as unbudded cells. Thus, a simple way of determining the growth state of a culture is to determine the frequency of budded cells in the microscope. Note that for some strains, the mother and daughter cells remain stuck together even though they have completed cytokinesis. In these cases, it is necessary to vortex or sonicate the culture to separate cells before microscopy. Many kinds of mutants also arrest in the cell cycle in a way that is diagnostic of their phenotype. For example, cells in which there is a defect in the mitotic spindle arrest as large budded cells, a point in the cycle that would normally correspond to mitosis. It is important to note that the arrest point, or terminal phenotype, of mutant cells can be morphologically distinct from any cell type seen in a normal culture. In the mitotic mutant above, the mother and daughter cells continue to grow at the arrest point until both are much larger than normal yeast cells.

Haploids versus Diploids

Haploid and diploid yeast cells (see Fig. 1) are morphologically similar but differ in several important ways. First, diploid cells are larger than haploid cells. Cytoplasmic volume increases with ploidy, and the diameter of a diploid cell is roughly 1.3 times that of a haploid cell. This difference can readily be seen when haploids and diploids are compared side by side. Because they are larger, diploid cells (or even tetraploids in some cases) are often used for fluorescence microscopy where the larger size helps to resolve small cellular structures. Second, diploid cells tend to have a more elongated, or ovoid, shape than round haploid cells. Third, diploids and haploids have different budding patterns. Yeast cells generally bud about 20 times before becoming senescent. Successive buds emerge from the surface of the mother cell in stereotyped patterns. Haploid cells bud in an axial pattern wherein each bud emerges adjacent to the site of the previous bud. Diploid cells bud in a polar pattern wherein successive buds can emerge from either end of the elongated mother cell. The history of a cell's budding pattern can be visualized by staining cells with Calcofluor, a fluorescent compound that binds to the rings of chitin that remain at old bud sites. These chitin rings are called bud scars, and we will use Calcofluor staining of haploid and diploid cells to visualize the axial and polar bud scar patterns.

Mating Cells

Yeast cells come in three mating types: *MAT***a** and *MAT*α; these two are able to mate with one another to yield a *MAT***a**/*MAT*α. *MAT***a**/*MAT*α cells cannot mate with cells of either mating type. Generally, *MAT***a** and *MAT*α strains will be haploid and *MAT***a**/*MAT*α strains will be diploid, although this is not always the case, and one should be careful to consider mating type independent of ploidy. The mating process between two cells begins with an exchange of pheromones that causes each of the cells to arrest in the cell cycle as unbudded cells and to induce the expression of proteins required for mat-

Figure 1. Examples of fluorescence microscopy. The upper panels show double labeling of actin filaments with rhodamine-phalloidin and chitin/bud scars with Calcofluor on a haploid vs. diploid cell. Note the differences in size and budding patterns. The middle panels show staining of mitochondria with $DiIC_5(3)$ on cells grown on a fermentable carbon source (glucose, *right*) vs. a nonfermentable carbon source (ethanol/glycerol, *left*). Note the elaboration of the mitochondria on the nonfermentable carbon source. The lower left panel shows the fluorescence of a GFP-Shs1p fusion protein thereby identifying the septin ring. The lower right panel shows a cell stained with FM4-64 thereby identifying the vacuolar membrane. (*See facing page.*)

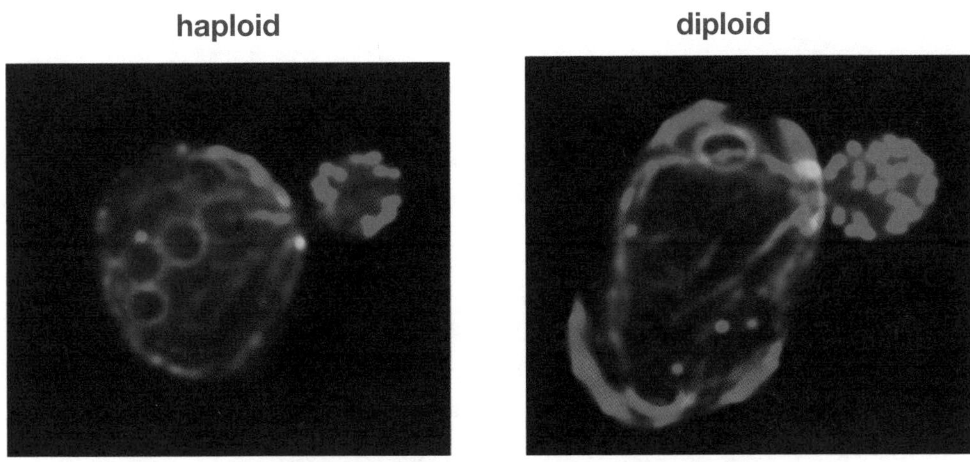

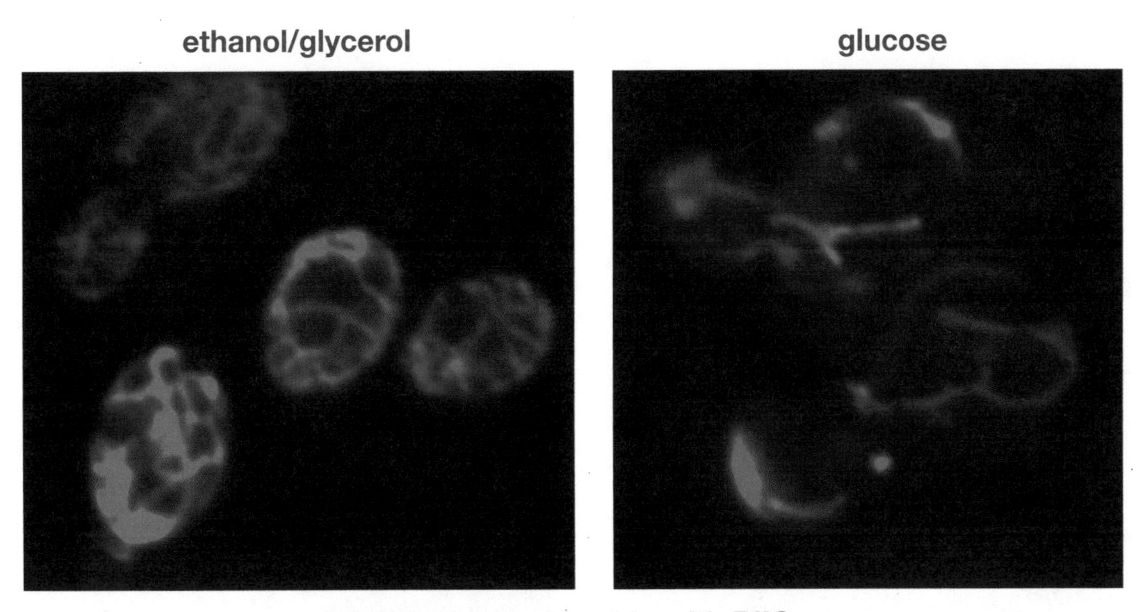

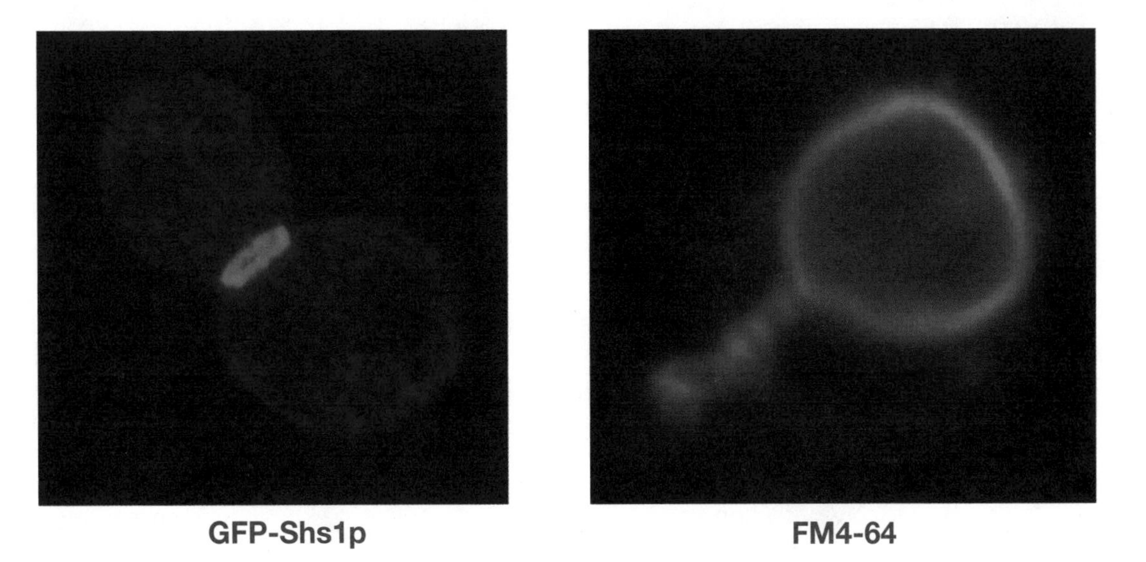

Figure 1. *(See facing page for legend.)*

ing. The pheromone also causes the cells to make a projection of new cell surface specialized for cell fusion. This projection is usually oriented toward the mating partner. Cells with a mating projection are called "shmoos" because of their resemblance to an Al Capp cartoon character from the 1940s. The shmooing cells join at the tips of their projections, their cytoplasms fuse, and their nuclei fuse to form a diploid *MAT***a**/*α* nucleus. The process of nuclear fusion is termed karyogamy. The newly formed diploid is termed a zygote and has a characteristic appearance that is particularly easy to identify when the first bud emerges. It is possible to isolate zygotes by micromanipulation, allowing for the isolation of diploid cells even in situations where there is no genetic selection for diploid formation. We will look at a population of mating cells to identify shmoos and zygotes.

Mitochondria

Yeast mitochondria contain a genome that encodes several proteins involved in oxidative phosphorylation, one ribosomal protein, and the rRNAs and tRNAs required for the mitochondrial translation apparatus. The vast majority of mitochondrial proteins are encoded by nuclear genes and imported into mitochondria from the cytoplasm. Thus, mutations in either the nuclear genome or the mitochondrial genome can affect mitochondrial function. Yeast cells with dysfunctional mitochondria cannot carry out oxidative phosphorylation and so must get all of their energy from fermentation. Such mutants grow more slowly than wild-type cells and are unable to grow on nonfermentable carbon sources such as lactate, glycerol, or ethanol. French scientists who first characterized mitochondrial mutants called this the "petite" phenotype. Petite strains form small milky-white colonies on fermentable carbon sources. This lack of colony pigmentation is most evident in an *ade2* background; formation of the red pigment that typifies *ade2* mutants requires oxidative phosphorylation. Diploid petite strains are also unable to sporulate, and it is wise to check a nonsporulating strain for the ability to grow on a nonfermentable carbon source before other potential causes of sporulation failure are examined. Petite mutants appear with high frequency in many common lab strains—for some strains, as many as 10% of the cells in a culture are petite. Although the petite phenotype can be a result of mutations in either the mitochondrial or nuclear genomes, the great majority of petites are the result of mitochondrial DNA mutations. The mitochondrial genome is given the designation "ρ," wild-type strains are ρ^+, strains with deleted versions of the mitochondrial genome (the most common type of mutation) are ρ^-, and strains lacking the mitochondrial genome entirely are ρ^o. A common misconception is that ρ^o strains lack mitochondria altogether. Several essential reactions take place within the mitochondrial membrane, and even in ρ^o strains, a diminished mitochondrial structure can be seen in the electron microscope. In this experiment, we use fluorescence microscopy to visualize the mitochondria and mitochondrial DNA in wild-type cells.

FLUORESCENCE MICROSCOPY

Because yeast cells have been increasingly used for experiments in cell biology, methods for determining the intracellular localization of gene products have been developed (Pringle et al. 1989), usually by adaptation of methods first used in animal cells. The association of gene products with known cellular structures has frequently led to important insights into the function of the genes involved. Moreover, examination of the morphologies of the nucleus and cytoskeleton gives precise information about the cell cycle stage of individual cells and has been used to stage cell division cycle mutants.

In this experiment, we use three different methods of identifying structures or proteins in yeast cells using fluorescence microscopy. The first is immunofluorescence, in which fixed cells are incubated first with primary antibodies against the protein of interest, and then with fluorescently labeled secondary antibodies directed against the primary antibodies. This layering of primary and secondary antibodies increases the potential fluorescent signal. We use immunofluorescence with antitubulin antibodies to visualize the microtubule cytoskeleton. This method has the advantage that it usually generates a strong fluorescence signal, but the cells must be fixed before observation and the cell wall must be removed to allow access of the antibodies. Care must be taken to avoid artifacts resulting from the cell preparation.

The second fluorescence microscopy method makes use of small molecules that are fluorescent and either bind to specific proteins in the cell or partition to certain organelles. We use 4′,6-diamidino-2-phenylindole (DAPI) to stain DNA; 1,1′-dipentyl-3,3,3′,3′-tetramethylindocarbocyanine iodide (DiIC$_5$(3)) (see Fig. 1) to stain mitochondria; rhodamine-phalloidin to stain the actin cytoskeleton; Calcofluor to stain bud scars; and FM4-64 (see Fig. 1) to stain vacuoles and endocytic vesicles. DAPI binds specifically to DNA and becomes more intensely fluorescent when bound. DiIC$_5$(3) is a fluorescent hydrophobic molecule that is specifically transported into mitochondria. Phalloidin is a toxin from the *Amanita phalloides* mushroom that binds to actin filaments; when labeled with rhodamine, it allows visualization of actin filaments. Calcofluor binds preferentially to the chitin of the cell wall concentrated in the bud scars. FM4-64 is a lipophilic styryl dye that is taken up by yeast cells via the endocytic pathway. It selectively stains the membranes of acidic compartments. DAPI and Calcofluor can be used on fixed or living cells, DiIC$_5$(3) and FM4-64 must be used on living cells, and rhodamine-phalloidin is usually used on fixed cells. Dyes that can be used to stain living cells are called vital stains.

The last and most recently developed method makes use of fluorescent proteins, such as GFP (see Fig. 1), a naturally fluorescent protein from the jellyfish *Aequoria victoria*. GFP is remarkable in that it retains its fluorescence when expressed in bacterial, fungal, plant, and animal cells, making it an ideal fluorescent marker protein. To identify the intracellular localization of a protein, the gene for that protein is fused to the GFP coding sequence, such that a fusion protein is created. Fusion to GFP does not usually affect function but this should always be confirmed by complementation. Expression of this fusion protein in cells allows visualization of the protein in living cells, and (usually) in fixed cells as well, provided that methanol is avoided in the fixation procedure. Although care must be taken to ensure that the fusion protein behaves like the wild-type protein, the GFP fusion method is extremely powerful because it allows dynamic behaviors to be observed in living cells. We look at cells expressing either GFP alone, or a fusion of GFP to *SHS1*; a protein component of the septin cystoskeleton; or a fusion of GFP to *AIP3/BUD6* a regulator of polarized actin assembly.

STRAINS

1-1	FY86	*MATα ura3-52 leu2Δ1 his3Δ200*
1-2	FY23	*MAT***a** *ura3-52 leu2Δ1 trp1Δ63*
1-3	FY23 x 86	*MAT***a***/MATα ura3-52/ura3-52 leu2Δ1/leu2Δ1 trp1Δ63/TRP1 HIS3/his3Δ200*
1-4	FY23 x 86	*MAT***a***/MATα ura3-52/ura3-52 leu2Δ1/leu2Δ1 trp1Δ63/TRP1 HIS3/his3Δ200* [pTD125]
1-5	FY23 x 86	*MAT***a***/MATα ura3-52/ura3-52 leu2Δ1/leu2Δ1 trp1Δ63/TRP1 HIS3/his3Δ200* [pDAb204]
1-6	FY23 x 86	*MAT***a***/MATα ura3-52/ura3-52 leu2Δ1/leu2Δ1 trp1Δ63/TRP1 HIS3/his3Δ200* [pTY20]

PLASMIDS

pTD125	YCp *URA3 GFP*
pDAb204	YCp *URA3 GFP-AIP3*
pTY20	YCp *URA3 GFP-SHS1*

PROCEDURE

SAFETY NOTES

Formaldehyde is toxic and a carcinogen. It is readily absorbed through the skin and is irritating to the eyes, skin, mucous membranes, and upper respiratory tract. Avoid breathing the vapors. Wear appropriate gloves and safety glasses, and always work in a chemical fume hood. Keep away from heat, sparks, and open flame.

DAPI is a possible carcinogen. It may be harmful if it is inhaled, swallowed, or absorbed through the skin. It may also cause irritation. Wear appropriate gloves, face mask, and safety glasses, and do not breathe the dust and vapors.

Day 1

You will be provided with cultures of strain 1-1 that were fixed with 4% formaldehyde during either log phase growth or stationary phase. Examine these using differential interference contrast (DIC) microscopy and count 100 cells of each culture, noting the numbers of unbudded, small budded, and large budded cells.

You will also be provided with a fixed mating mixture of 1-1 and 1-2 cells. Examine this culture using DIC microscopy and identify shmoos and zygotes, based on their distinctive morphology.

Compare the morphologies of haploid 1-1 and diploid 1-3 cells, looking for differences in the size and shape of the cells. In the morning, start overnight cultures of

strain 1-1 in 5 ml of YPD at 30°C
strain 1-3 in 25 ml of YPD at 30°C
strain 1-3 in 5 ml of YPEG (same as YPD except 3% ethanol and 3% glycerol replace glucose as the carbon source) at 30°C
strains 1-4, 1-5, and 1-6 in 5 ml of SC-uracil at 30°C

Day 2

Mitochondria. Dilute the 1-3 YPD and YPEG cultures from yesterday into YPD and YPEG so that they go through at least three doublings to achieve a density of approximately 1×10^7 to 2×10^7/ml (midlog). We refer to this technique as "dilute back." Stain with 50–100 ng/ml $DiIC_5(3)$ for 5–10 minutes and examine under the fluorescent microscope using the rhodamine filter set.

Bud scars. Dilute back the 1-1 and 1-3 YPD cultures from yesterday if they are overgrown and grow to early log phase (1×10^7 to 2×10^7/ml) in 5 ml, passing through at least three doublings. Stain with Calcofluor according to Techniques and Protocols #11D, Calcofluor Staining of Chitin and Bud Scars.

Vacuoles. Dilute back the 1-3 YPD culture from yesterday if it is overgrown and grow to early log phase (1×10^7 to 2×10^7/ml) in 5 ml, passing through at least three doublings. Stain with FM4-64 according to Techniques and Protocols #11C, Visualizing Vacuoles and Endocytic Compartments with FM4-64.

Septin and polarisome visualization. Dilute back the 1-4, 1-5, and 1-6 SC-uracil cultures from yesterday if they are overgrown and grow to early log phase (1×10^7 to 2×10^7/ml) in 5 ml, passing through at least three doublings. Examine patterns of GFP fluorescence under the fluorescent microscope.

Actin and tubulin staining. In the morning, dilute back the 1-3 YPD culture from yesterday to grow to 2×10^7/ml in 50 ml of YPD, passing through at least three doublings. Fix by adding 34 ml of 10% EM-grade formaldehyde (Polysciences) and incubate for 10 minutes. Split the fixed cells into equal halves and follow the fixation procedures outlined in the rhodamine-phalloidin staining procedure described in

Techniques and Protocols #13, Actin Staining in Fixed Cells, and the immunofluorescence protocol described in Techniques and Protocols #12, Yeast Immunofluorescence. Store the fixed cells overnight at 4°C.

Day 3

Actin staining. Complete the rhodamine-phalloidin staining procedure described in Techniques and Protocols #13. Examine the cells using the rhodamine filter set on the fluorescence microscope.

Antitubulin immunofluorescence. Complete the immunofluorecence staining procedure described in Techniques and Protocols #12. Use the YOL1/34 antitubulin antibody at 1:500 and the goat anti-rat FITC secondary antibody at 1:200. Examine the cells using the fluorescein filter set under the fluorescence microscope. Note that since the mounting solution contains DAPI, nuclear DNA staining can also be viewed in the DAPI channel.

MATERIALS

Day 1

Fixed cultures of 1-1 and 1-3, and fixed culture of 1-1 mating to 1-2
30 ml of YPD
5 ml of YPEG
15 ml of SC-uracil
Slides and coverslips

Day 2

70 ml of YPD
5 ml of YPEG
1 µl of 2.5 mg/ml stock of $DiIC_5(3)$ in ethanol
1 ml of 1 mg/ml stock of Calcofluor in H_2O
2 µl of 8 mM FM4-64 in H_2O (Molecular Probes #T-3166)
15 ml of SC-uracil
42 ml of 10% EM-grade formaldehyde (Polysciences)
17 ml of PBS
17 ml of 40 mM KPO_4 (pH 6.5)
500 µM $MgCl_2$
6 ml of 1.2 M sorbitol
40 mM KPO_4 (pH 6.5)
500 µM $MgCl_2$
Slides and coverslips

Day 3

20 µl of rhodamine-phalloidin, 6.6 µM in MeOH (Molecular Probes #R-415)

5 ml of PBS
Mounting solution
30 μl of Zymolyase 100T, 10 mg/ml in 1.2 M sorbitol
40 mM KPO_4 (pH 6.5)
500 μM $MgCl_2$
5 ml of 1.2 M sorbitol
40 mM KPO_4 (pH 6.5)
500 μM $MgCl_2$
1 Teflon-faced multiwell slide
50 μl of 1% polylysine stock in H_2O
H_2O, clean, sterile, and spun at high speed to remove particulates
2 coplin jars, one containing dry ice cold acetone, the other dry ice cold methanol
5 ml of PBS (pH 7.4)
0.5% of BSA
0.5% of ovalbumin
200 μl of 1:500 diln
YOL1/34 antitubulin antibody in PBS (pH 7.4)
0.5% of BSA
0.5% of ovalbumin
200 μl of goat anti-rat FITC secondary antibody, diluted 1:200 in PBS (pH 7.4)
0.5% of BSA
0.5% of ovalbumin
Large coverslip
Sally Hansen Hard as Nails

REFERENCE

Pringle J.R., Preston R.A., Adams A.E.M., Stearns T., Drubin D.G., Haarer B.K., and Jones E.W. 1989. Fluorescence microscopy methods for yeast. *Methods Cell Biol.* **31:** 357–435.

EXPERIMENT II

Isolation and Characterization of Auxotrophic, Temperature-sensitive, and Osmotic-sensitive Mutants

Since spontaneous mutation frequencies are low, yeast is usually treated with such mutagens as ultraviolet (UV) radiation, nitrous acid, ethylmethanesulfonate (EMS), diethyl sulfate, and 1-methyl-nitro-nitrosoguanidine to enhance the frequency of mutants. These mutagens are remarkably efficient and can induce mutations at a rate of 5×10^{-4} to 1×10^{-2}/gene without substantial killing. Even though there are known methods to increase the proportion of mutants by killing off the nonmutants with nystatin and other agents, it is usually unnecessary to use selective means to obtain reasonable yields of mutants (Snow 1966; Thouvenot and Bourgeois 1971; Henry et al. 1975; Walton et al. 1979). In this experiment, auxotrophic, temperature-sensitive, and osmotic-sensitive mutants are isolated from EMS-treated yeast.

Auxotrophic mutants have been invaluable for the elucidation of biochemical pathways as well as for the study of the relationship between enzyme structure and function (Lingens and Oltmanns 1964, 1966; Lindegren et al. 1965). Studies on the intermediates accumulated by amino acid auxotrophs have facilitated the unraveling of biochemical pathways.

Studies on temperature-sensitive mutants make it possible to identify and study essential genes (Hartwell 1967; Pringle and Hartwell 1981). Many yeast genes specify proteins that participate in indispensable functions (i.e., RNA polymerases, tRNA synthetases, etc.). Mutations that completely destroy the activity of these proteins are lethal, and those that give only partial activity are genetically and biochemically difficult to work with. A mutation that affects the structure of one of these indispensable proteins, such that it can function at low temperature but not at high temperature, is much more useful. A strain carrying such a mutation grows at normal or nearly normal rates at low temperature but does not grow on any medium at elevated temperatures. This phenotype distinguishes the temperature-sensitive mutation in a vital function from a supplementable temperature-sensitive mutant (i.e., one whose defect is in the biosynthesis of an amino acid).

All eukaryotic cells must be able to adapt to changing environmental conditions such as temperature, nutritional sources, and alterations in external osmolarity. Yeast in the wild live in a hostile and dynamic environment. For example, in vineyards the osmolarity of their surroundings can change dramatically from a rain drop to a concentrated sugar solution of a grape. For cells to maintain proper water activity and bring water into the cell, they must maintain an internal osmolarity slightly above that of their external environment. For these reasons, yeast have evolved finely tuned systems to sense osmotic imbalances across their plasma membrane and initiate adaptive changes in cell architecture (Brewster and Gustin 1994) and gene expression (O'Rourke and Herskowitz 2004), culminating in the production of high concentrations of internal glycerol (Edgley and Brown 1983), the main osmo-balancing agent of yeast cells. Studies on osmotic-sensitive mutants of yeast have uncovered a conserved MAP kinase pathway (Brewster et al. 1993) that functions in all eukaryotic cells to mediate adaptation to stress conditions: the high-osmolarity growth or HOG pathway (for a comprehensive review of osmo adaptation in yeast, see Hohmann 2002).

In certain cases, it is valuable to be able to select directly for mutations in particular genes. For example, sometimes it is convenient to introduce auxotrophic markers into strains without having to put them through a genetic cross. Two particularly easy direct selections for auxotrophic mutations are the use of α-aminoadipate (αAA) for selection of *lys2* and *lys5* mutations (Chattoo et al. 1979; Zaret and Sherman 1985) and the use of 5-fluoro-orotic acid (5-FOA) for selection of *ura3* and *ura5* mutations (Boeke et al. 1986). In this exercise, we explore the frequencies and phenotypes of 5-FOA-resistant mutants.

The mutagenesis described here involves treatment of wild-type yeast strains with EMS. Half of the class will mutagenize a *MAT***a** strain, and half a *MAT*α strain. After mutagenesis, the strains will be diluted and plated onto complete medium plates at a concentration of about 200 cells/plate. After these cells have grown into colonies, they will be transferred to various media by replica-plating. Temperature-sensitive mutants will be detected by comparing pairs of plates that were incubated at room temperature (about 23°C) and 37°C. Osmotic-sensitive mutants will be detected by comparing plates containing 1.2 M sorbitol with the controls. Auxotrophic mutants will be detected by lack of growth on a minimal medium that contains glucose, potassium phosphate, ammonium sulfate, a few vitamins, salts, and trace metals. The specific requirements can be determined by testing the colony from the original YPD plate on various types of synthetic media.

To determine the specific auxotrophic defects of mutants able to grow on complete but not minimal defined medium, putative mutants will be transferred to nine different minimal (SD) medium plates supplemented with pools of various amino acids, purines, and pyrimidines, and other metabolites (see below). From the pattern of growth of a particular strain on these plates, it will be possible to identify the specific requirement of the mutant strain.

Pools	#1	#2	#3	#4	#5
#6	adenine	guanine	cysteine	methionine	uracil
#7	histidine	leucine	isoleucine	valine	lysine
#8	phenylalanine	tyrosine	tryptophan	threonine	proline
#9	glutamate	serine	alanine	aspartate	arginine

Generally, a colony will respond on a plate containing one of the pools from 1 to 5 and on another plate containing one of the pools from 6 to 9, thus allowing direct identification of a single growth factor requirement. For example, a colony growing on pools 1 and 7 requires histidine, whereas a colony growing on pools 3 and 8 requires tryptophan. If a colony grows on only one of the nine pools, it requires more than one of the nutrients in that pool. A mutant blocked early in the pathway of aromatic amino acid biosynthesis will grow only on pool 8.

STRAINS

2-1	S288C	*MAT*α *mal gal2*
2-2	D665-1A	*MAT***a**

PROCEDURE

SAFETY NOTES

Ethylmethanesulfonate (EMS) is a volatile organic solvent that is a mutagen and carcinogen. It is harmful if inhaled, ingested, or absorbed through the skin. Discard supernatants and washes containing EMS in a beaker containing 50% sodium thiosulfate. Decontaminate all material that has come in contact with EMS by treatment in a large volume of 10% (w/v) sodium thiosulfate. Use extreme caution when handling. When using undiluted EMS, wear appropriate protective gloves and use in a chemical fume hood. Store EMS in the cold. DO NOT mouth-pipette EMS. Pipettes used with undiluted EMS should not be too warm; chill them in the refrigerator before use to minimize the volatility of EMS. All glassware coming in contact with EMS should be immersed in a large beaker of 1 N NaOH or laboratory bleach before recycling or disposal.

Day 1

Inoculate 5 ml of YPD with one of the strains above (half of the class will use the *MAT*α strain 2-1, and the remainder of the class will use the *MAT***a** strain 2-2). Grow overnight at 30°C.

Day 2

Mutagenize your cells with EMS using the method described in Techniques and Protocols #21, EMS Mutagenesis. On completion of the mutagenesis protocol, you will have one tube of mutagenized cells and one nonmutagenized control, and you will take the following three steps with these cells:

1. Spread the cells directly on αAA medium and 5-FOA medium to select for *lys2* and *ura3* mutants.
2. Grow the cells in YPD overnight and then spread on 5-FOA medium.
3. Dilute and spread the cells on YPD plates for colonies to be screened for other mutations.

First, spread 0.1 ml of the mutagenized cells directly on a 5-FOA plate. In addition, as a control, plate 0.1 ml of the washed, nonmutagenized cells onto a separate 5-FOA plate. The frequencies of colonies arising from the mutagenized and nonmutagenized cultures will also serve to monitor the frequency of mutagenesis caused by the EMS treatment.

Second, inoculate 0.1 ml of each cell suspension (mutagenized and nonmutagenized) into 1 ml of YPD broth. Grow overnight at 30°C.

Third, to achieve the goal of obtaining 200 colonies/YPD plate, dilute your EMS mutagenized cells to a concentration of 2000 cells/ml (based on the hemocytometer count, above, and assuming no loss of cells in the mutagenesis procedure). This should be about a 1:100,000 dilution, but may need to be adjusted depending on the initial concentration of cells used. Spread 0.1, 0.2, and 0.4 ml each on separate YPD plates, using ten plates for each of the three different volumes plated. Incubate all of the plates for 4 days at room temperature (23°C).

Day 3

Wash the 1-ml YPD overnight cultures from day 2 with H_2O and resuspend in 1 ml of H_2O. Spread 0.1 ml onto a 5-FOA plate. Incubate at 30°C.

Day 6

Examine and count the colonies on your YPD plates and estimate the survival after EMS. Choose ten or more YPD plates containing approximately 200 colonies/plate for the isolation of (1) auxotrophic, (2) temperature-sensitive, and (3) osmotic-sensitive mutants. Transfer the colonies from each of the YPD plates by replica-plating onto one SD plate, one SC plate, one YPD + 1.2 M sorbitol plate, and three YPD plates. Record the number of colonies transferred. Be sure that each plate is numbered and has an orientation symbol on the back. For detection of auxotrophic mutants, incubate the SC and SD plates for 1 day at 30°C. For temperature-sensitive mutants, incubate one of the YPD replicas per set for 2 days at 37°C, and one for 2 days at room temperature (23°C).

For osmotic-sensitive mutants, incubate one set of YPD plates and the YPD + 1.2 M sorbitol plates for 2 days at 30°C.

ura3 *mutants.* Place the 5-FOA plates from day 2 at 4°C.

Day 7

Auxotrophic mutants. Compare each of the ten SD plates with the SC plates to identify candidate auxotrophic mutants. There should be 5% such colonies. Try to identify at least 20 candidate colonies. We use a multipronged inoculating device (a frogger) to transfer cells from each of these colonies onto a series of plates designed to reveal the specific auxotrophic deficiencies of each candidate. To "frog," place 100 µl of sterile H_2O in the wells of a microtiter dish; in each well, using a sterile toothpick, suspend an aliquot of cells from one of your selected colonies. Take care to add the same amount of inoculum from each colony selected and to suspend it well. Include strain 2-1 or 2-2 also. Transfer a droplet from each well onto the appropriate plates by lowering the prongs of the frogger into the wells, lifting it out quickly, then touching it to the surface of a fresh plate. Using frogging, inoculate the nine pool plates, an SC, and an SD plate with cells from your candidate auxotrophs. Incubate overnight at 30°C.

ura3 *mutants.* Count the number of colonies arising on the 5-FOA plates spread on days 2 and 3. Record the data in the table below. Estimate the frequency of each type of mutant before and after mutagenesis and before and after outgrowth following mutagenesis, taking 2×10^8 cells/ml as the saturation density of your YPD overnight cultures. Purify four colonies from a 5-FOA plate on another YPD for subsequent analysis of their Ura phenotype. Incubate the plates at 30°C. Did the frequency of *ura3* mutants increase after mutagenesis? Why might growth after mutagenesis affect the recovery of each type of mutant?

Day 8

Medium treatment	5-FOA	
	No outgrowth	Outgrowth
Number of colonies per plate		

Auxotrophic mutants. Record the growth response of your candidates. Assign each candidate a name that corresponds to its auxotrophy (e.g., leu1, leu2, etc.). Record the names of your auxotrophs in the appropriate categories of the classwide listing of auxotrophs. Also, label the corresponding auxotrophs on your SC plates from day 7. You will be assigned a specific category of auxotrophs with which to perform a complementation test. Collect all the auxotrophs of that category from your classmates. Make two streak master YPD plates of your category of auxotrophs: one for the *MAT***a** strains and one for the *MATα* strains. Include strain 2-1 on the *MATα* plate and strain 2-2 on the *MAT***a** plate. Each streak plate should have up to 11 parallel streaks that continue across the entire surface of the plate (see Appendix D, Templates for Making Streak Plates, for a template). The top and bottom streaks should be about 2 cm from the top and bottom of the plate (respectively). Incubate overnight at 30°C, then refrigerate.

Osmotic-sensitive mutants. Compare the YPD + 1.2 M sorbitol plates with the YPD controls. Pick approximately 20 mutants that fail to grow on the high-osmolarity plates but grow well on the YPD plates and transfer the mutants by frogging onto one YPD plate and one YPD + 1.2 M sorbitol plate. Include strain 2-1 or 2-2 also. Incubate the two plates for 2 days at 30°C.

Temperature-sensitive mutants. Compare the 37°C plate with the 23°C plate. Identify up to 20 colonies that failed to grow at the high temperature. Use the frogging method to inoculate cells from these Ts^- candidates, as well as either the 2-1 or 2-2 control, onto two YPD plates. Also frog onto YPD containing 30% sucrose. This plate will allow you to identify those temperature-sensitive mutants whose defects cause osmotic lysis on the YPD plate or who are suppressed by high osmotic conditions (osmotic-remedial mutants). Incubate one YPD plate and the YPD + sucrose plate at 37°C, and the other YPD plate for 2 days at room temperature.

ura3 ***mutants.*** Streak the purified 5-FOA^R colonies onto SD and SD + uracil plates to test the Ura phenotype.

Day 10

Auxotrophic mutants. Sequentially replica-plate the *MAT***a** and *MATα* streak master plates onto one velvet, such that the streaks are perpendicular. Print from this velvet onto a fresh YPD plate. Incubate at 30°C.

Osmotic-sensitive mutants. Record the growth response and assign a number to each mutant. Make two identical streak master YPD plates with up to ten of your mutants (plus the 2-1 or 2-2 control). Incubate overnight at 30°C.

Temperature-sensitive mutants. Record the growth of your mutants at 23°C and 37°C. Assign a number to each mutant. Make two identical streak master plates of your mutants (plus the 2-1 or 2-2 control) on a pair of YPD plates. Incubate for 2 days at room temperature.

ura3 ***mutants.*** Record the growth of the *ura3* mutants from day 9.

Day 11

Auxotrophic mutants. Replica-plate the cross-streaked mutants to one SC and one SD plate. Incubate at 30°C.

Osmotic-sensitive mutants. Exchange one of your streak plates for a streak plate of osmotic-sensitive mutants of the opposite mating type. Use replica-plating to create a cross-streaked YPD plate. Incubate overnight at 30°C.

Day 12

Auxotrophic mutants. Score your complementation test. How many cases of non-complementation did you observe? Explore the *Saccharomyces* Genome Database (http://pathway.yeastgenome.org/biocyc/) to estimate the number of genes that can be mutated to yield the auxotrophic phenotype that you are studying.

Osmotic-sensitive mutants. Replica-plate your cross-streaked mutants to one YPD plate and one YPD + 1.2 M sorbitol plate and incubate both plates for 2 days at 30°C. A diploid is expected to have normal osmotic sensitivity if it was formed from two recessive Osm^- strains with mutations at different loci.

Temperature-sensitive mutants. Exchange one of your streak plates for a streak plate of Ts^- mutants of the opposite mating type. Use replica-plating to create a cross-stamped YPD plate; incubate for 2 days at 23°C.

Day 14

Temperature-sensitive mutants. Replica-plate your cross-streaked mutants to 2 YPD plates; for 2 days incubate one at 37°C and the other at 23°C.

Osmotic-sensitive mutants. Score the complementation tests. How many complementation groups did you get and why do you think there are so few?

Day 16

Temperature-sensitive mutants. Score the complementation tests.

MATERIALS

Note: Amounts provided are those required for each pair.

Day 1

1 culture tube, containing 5 ml of YPD

Day 2

10 ml of sterile distilled H_2O
5 ml of sterile 0.1 M sodium phosphate buffer (pH 7)
EMS (ethylmethanesulfonic acid ester; Sigma M0880)
1 ml of sterile 5% sodium thiosulfate (w/v)
2 culture tubes, each containing 1 ml of YPD
2 5-FOA plates
30 YPD plates

Day 3 10 ml of sterile H_2O
2 5-FOA plates

Day 6 10 SC plates
10 SD plates
10 YPD + 1.2 M sorbitol plates
30 YPD plates
10 sterile velveteen squares

Day 7 1 YPD plate
1 each of the 9 pool plates
1 SD plate
1 SC plate
10 ml of sterile H_2O
2 sterile microtiter dishes
95% ethanol (for flaming frogger)

Day 8 5 YPD plates
1 YPD + 1.2 M sorbitol plate
1 SD plate
1 SD + uracil plate
1 YPD plate containing 30% sucrose
3 sterile microtiter dishes
10 ml of sterile H_2O
95% ethanol (for flaming frogger)

Day 10 5 YPD plates
1 sterile velveteen square

Day 11 1 YPD plate
1 SD plate
1 SC plate
2 sterile velveteen squares

Day 12 2 YPD plates
1 YPD + 1.2 M sorbitol plate
2 sterile velveteen squares

Day 14 2 YPD plates
1 sterile velveteen square

REFERENCES

Auxotrophic Mutants

Lindegren G., Hwang, L.Y., Oshima Y., and Lindegren C. 1965. Genetical mutants induced by ethyl methanesulfonate in *Saccharomyces*. *Can. J. Genet. Cytol.* **7:** 491–499.

Lingens F. and Oltmanns O. 1964. Erzeugung und untersuchung Biochemischer and Mangelmutanten von *Saccharomyces cerevisiae*. *Z. Naturforsch.* **19B:** 1058–1065.

———. 1966. Uber die Mutagene Wirkung von 1-nitroso-3-nitro-1-methyl-guanidin (NNMG) und *Saccharomyces cerevisiae*. *Z. Naturforsch.* **21B:** 660–663.

Temperature-sensitive Mutants

Hartwell L.H. 1967. Macromolecule synthesis in temperature-sensitive mutants of yeast. *J. Bacteriol.* **93:** 1662–1670.

Pringle J.R. and Hartwell L.H. 1981. The *Saccharomyces cerevisiae* cell cycle. In *The molecular biology of the yeast* Saccharomyces: *Life cycle and inheritance* (ed. J.N. Strathern et al.), pp. 97–142. Cold Spring Harbor Laboratory, Cold Spring Harbor, New York.

Osmotic-sensitive Mutants

Brewster J.L. and Gustin M.C. 1994. Positioning of cell growth and division after osmotic stress requires a MAP kinase pathway. *Yeast* **10:** 425–439.

Brewster J.L., de Valoir T., Dwyer N.D., Winter E., and Gustin M.C. 1993. An osmosensing signal transduction pathway in yeast. *Science* **259:** 1760–1763.

Edgley M. and Brown A.D. 1983. Yeast water relations: Physiological changes induced by solute stress in *Saccharomyces cerevisiae* and *Saccharomyces rouxii*. *J. Gen. Micro.* **129:** 3453–3463.

Hohmann S. 2002. Osmotic stress signaling and osmoadaptation in yeasts. *Microbiol. Mol. Biol. Rev.* **66:** 300–372.

O'Rourke S.M. and Herskowitz I. 2004. Unique and redundant roles for HOG MAPK pathway components as revealed by whole-genome expression analysis. *Mol. Biol. Cell.* **15:** 532–542.

Selections for Auxotrophs

Boeke J.D., LaCroute F., and Fink G.R. 1986. A positive selection for mutants lacking orotidine-5′-phosphate decarboxylase activity in yeast; 5′-fluoro-orotic acid resistance. *Mol. Gen. Genet.* **197:** 345–346.

Chattoo B.B., Sherman, F. Azubalis D.A., Fjellstedt T.A., Mehnert D., and Ogur M. 1979. Selection of *lys2* mutants of the yeast *Saccharomyces cerevisiae* by the utilization of α-aminoadipate. *Genetics* **93:** 51–65.

Zaret K.S. and Sherman F. 1985. α-Aminoadipate as a primary nitrogen source for *Saccharomyces cerevisiae* mutants of yeast. *J. Bacteriol.* **162:** 579–583.

Enrichment Methods

Henry S.A., Donahue T.F., and Culbertson M.R. 1975. Selection of spontaneous mutants by inositol starvation in *Saccharomyces cerevisiae*. *Mol. Gen. Genet.* **143:** 5–11.

Snow R. 1966. An enrichment method for auxotrophic yeast mutants using the antibiotic "nystatin." *Nature* **211:** 206–207.

Thouvenot D.R. and Bourgeois C.M. 1971. Optimisation de la selection de mutants de *Saccharomyces cerevisiae* par la nystatine. *Ann. Inst. Pasteur* **120:** 617–625.

Walton B.F., Carter B.L.A., and Pringle J.R. 1979. An enrichment method for temperature-sensitive and auxotrophic mutants of yeast. *Mol. Gen. Genet.* **171:** 111–114.

EXPERIMENT III

Meiotic Mapping

The events of meiosis make it possible to deduce information about the positional relationships of genetic markers. For many years, meiotic mapping has been of great use in constructing genetic maps (Mortimer and Hawthorne 1966). Mortimer and Hawthorne (1969), Fincham et al. (1979), and Mortimer and Schild (1981) give general information concerning meiotic mapping and tetrad analysis. At present, genetic mapping techniques are instrumental in monitoring the manipulations of the genome, performed using molecular genetic techniques. The four spores in an ascus are the products of a single meiotic event, and the genetic analysis of these tetrads can provide linkage relationships of genes present in the heterozygous condition. Is is also possible to map a gene relative to its centromere if scorable alleles of known centromere-linked genes are present in the hybrid. Although the isolation of four spores from an ascus is a skill acquired only with considerable practice, tetrad analysis is useful not only for linkage studies, but also for constructing strains necessary for genetic and biochemical experiments. For many reasons, it is proper laboratory practice to make knockout alleles in diploid strains and confirm insertion of the allele by 2:2 segregation among meiotic tetrads.

In a cross of two haploids, *AB* x *ab* (*A* and *B* are two different genes, and uppercase vs. lowercase letters reflect distinguishable alleles), the segregation of the markers can yield three types of tetrads. The three classes of tetrads—parental ditype (PD), nonparental ditype (NPD), and tetratype (T)—from a diploid that is heterozygous for two markers, *AB* x *ab*, are shown in the table below.

	PD		NPD		T
	AB		*aB*		*AB*
	AB		*aB*		*Ab*
	ab		*Ab*		*ab*
	ab		*Ab*		*aB*
Random assortment	1	:	1	:	4
Linkage	>1	:	<1		
Centromere linkage	1	:	1	:	<4

LINKED MARKERS

The ratio of the three types of tetrads that is observed for a pair of markers is a function of the relative map positions of the markers. If the markers *A* and *B* segregate at random with respect to one another, the PD, NPD, and T patterns will be observed in a 1:1:4 ratio. However, if they are linked, they will yield different ratios that can be used to deduce their map distances. Figure 1 shows the outcome of zero, one, or two crossovers between markers *A* and *B* on the same chromosome. When there are no crossovers, a parental ditype is always formed. A single crossover always yields a tetratype. There are four possible types of double crossovers involving two, three, or four strands (chromatids). Two-strand double crossovers yield a PD, the two types of three-strand double crossovers both yield tetratypes, and a four-strand double crossover yields an NPD. Meioses in which more than two crossovers occur between *A* and *B* can result in any of the three types of tetrads, depending on which strands are involved. The probability that a crossover will occur between two markers is approximately proportional to the physical distance between them. Thus, markers that are closely linked yield predominantly PD tetrads and few NPD tetrads. In contrast, markers that are unlinked yield equal numbers of PD and NPD tetrads.

The distinctive segregation patterns exhibited by markers among which there have been zero, one, or two crossovers allow the formulation of a mapping function (Perkins

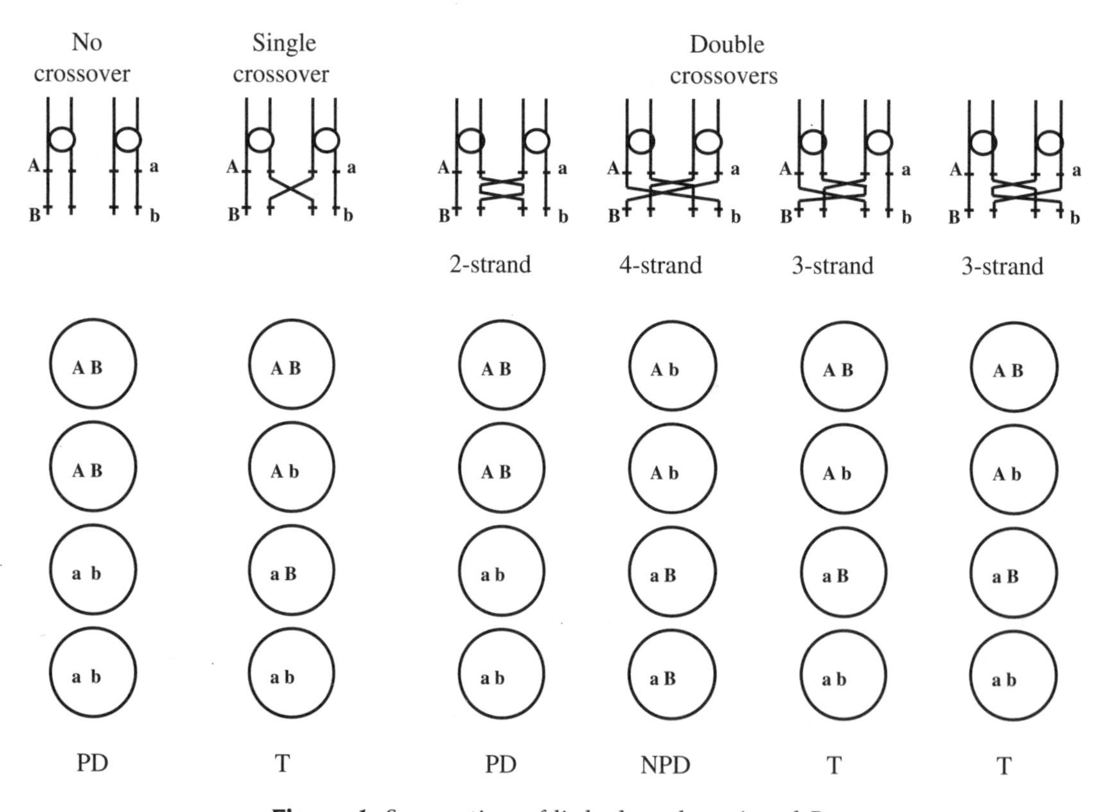

Figure 1. Segregation of linked markers *A* and *B*.

1949). For genetic intervals that experience at most two crossovers, the NPDs provide an accurate indicator of the frequency of double crossovers (DCOs), which in turn allows us to deduce the incidence of single crossovers (SCOs). The mapping function is an indicator of the average number of meiotic crossovers per chromatid between markers *A* and *B*. With a single crossover, half of the chromatids in the tetrad become recombinant, whereas double crossovers result in an average of one crossover per chromatid (although the crossovers do not always occur on all four chromatids; Fig. 1). The mapping function can therefore be expressed as

$$100 \times \frac{1/2(\text{SCO tetrads}) + (\text{DCO tetrads})}{\text{Total tetrads}}$$

This function gives us the number of centimorgans (cM) among the markers. To estimate the total number of tetrads in which a DCO occurred in our interval, we take advantage of the fact that one fourth of the DCO tetrads yields NPD tetrads (Fig. 1). With this information we can estimate the total number of DCO tetrads, assuming no chromatid interference, to be 4 (NPD tetrads). SCO tetrads are always tetratypes (Fig. 1), but tetratypes are also generated by DCO tetrads. To estimate the number of SCO tetrads, we subtract the contribution of DCO tetrads to the pool of tetratype tetrads. Because DCOs yield two Ts for each NPD (Fig. 1), this is 2 (NPD). Placing these values into the equation above gives us the following mapping function:

$$100 \times \frac{1/2(\text{T} - 2[\text{NPD}]) + 4(\text{NPD})}{\text{Total tetrads}} = 100 \times \frac{1/2\text{T} + 3(\text{NPD})}{\text{Total tetrads}}$$

UNLINKED MARKERS

A number of assumptions concerning interference must be made to determine map distances between large intervals. In addition, for larger intervals, we cannot tell whether an NPD tetrad is the consequence of two, three, or even more crossovers between *A* and *B*. The mapping function above is based on the assumption that there are at most two crossovers in the interval per meiosis, which therefore results in underestimates of the true genetic distance for intervals that experience multiple crossovers per meiosis. The only accurate way of measuring long intervals is by the summation of shorter intervals.

A special case of unlinked markers exists for those that are on different chromosomes, but close to their respective centromeres. Such situations yield equal numbers of PD and NPD tetrads (remember, PD = NPD is the hallmark of unlinked markers), but in this case instead of observing the PD:NPD:T ratio of 1:1:4, characteristic of random assortment, there is a reduction in the proportion of T asci. All tetratype asci require a crossover between the centromere and one of the markers.

The distance from a gene to its centromere can be estimated by determining the frequency of second-division segregation (SDS), explained below. First-division segregation (FDS) is the term applied to the movement of homologous centromeres to opposite poles at meiosis I (Fig. 2A). Heterozygous genes that map adjacent to the centromeres, such as *TRP1/trp1* (Fig. 2A), also show this segregation pattern in most meioses. Only when there is a crossover in the small interval between the *TRP1* locus and the nearby centromere do the *TRP1* and *trp1* alleles migrate away from one another at meiosis II, exhibiting SDS (Fig. 2B). For *TRP1*, this occurs in only about 1% of meioses, making the segregation of *TRP1* a good indicator of centromere behavior in meiosis. It is possible to determine the frequency of SDS for a gene of interest by comparing its segregation to the segregation of *TRP1*. If we perform tetrad analysis with a strain heterozygous for our gene of interest and *TRP1* (*A, TRP1* x *a, trp1*), PD and NPD tetrads will indicate instances of FDS of gene *A* (Fig. 3A) and T tetrads will indicate SDS (Fig. 3B). The map distance between gene *A* and its centromere can be approximated as

$$100 \times \frac{\text{(Tetratype tetrads)}}{2\ \text{(Total tetrads)}}$$

This mapping function is based on the assumptions that (1) T tetrads are the result of

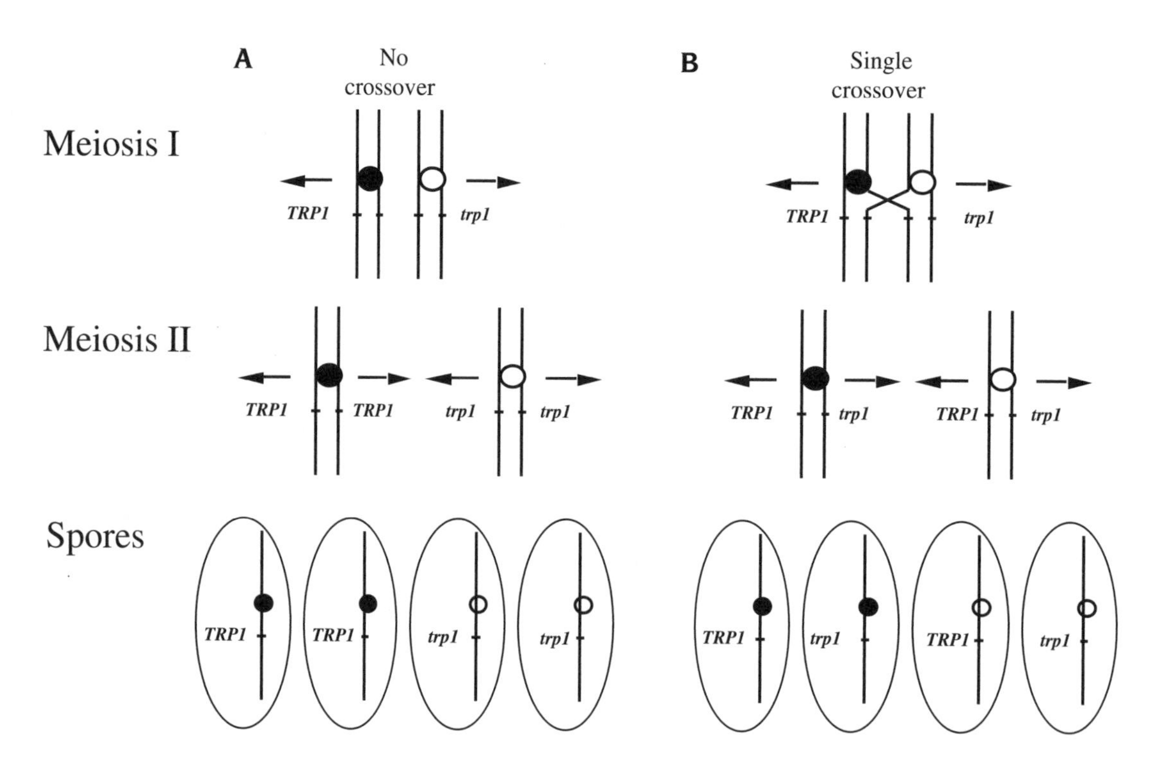

Figure 2. First- and second-division segregation. The gene, *TRP1*, is located close to the centromere of chromosome *IV*. In most meioses, these are no crossovers between *TRP1* and *CENIV*. In meioses with no crossovers in the *TRP1* to *CENIV* interval (*A*), heterozygous *TRP1* and *trp1* alleles migrate away from one another at the first meiotic division. This is called first-division segregation. In a fraction of meioses, there will be a crossover between the *TRP1* locus and *CENIV* (*B*). The result is that each homolog carries a *TRP1* allele on one chromatid and a *trp1* allele on the other. In this situation, the *TRP1* and *trp1* alleles do not migrate away from one another until meiosis II. This is called second-division segregation.

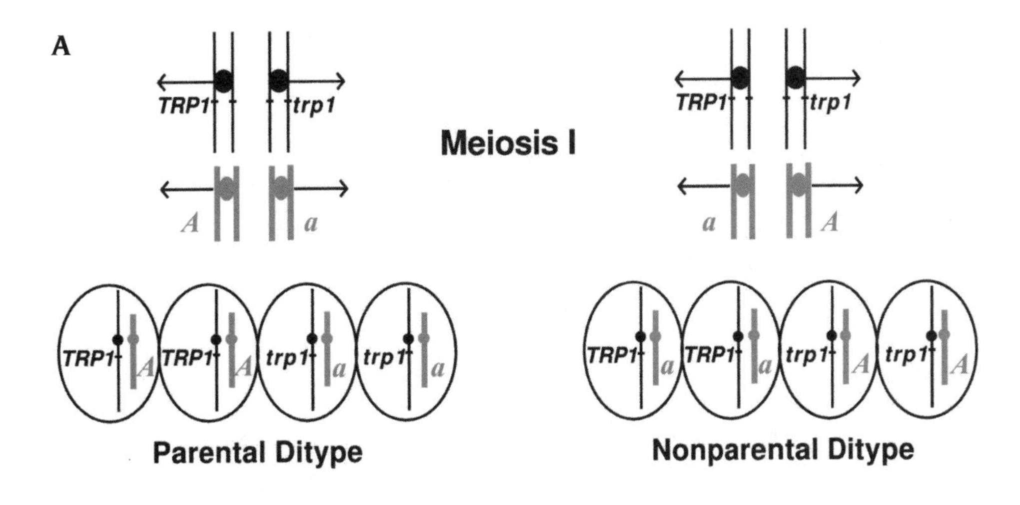

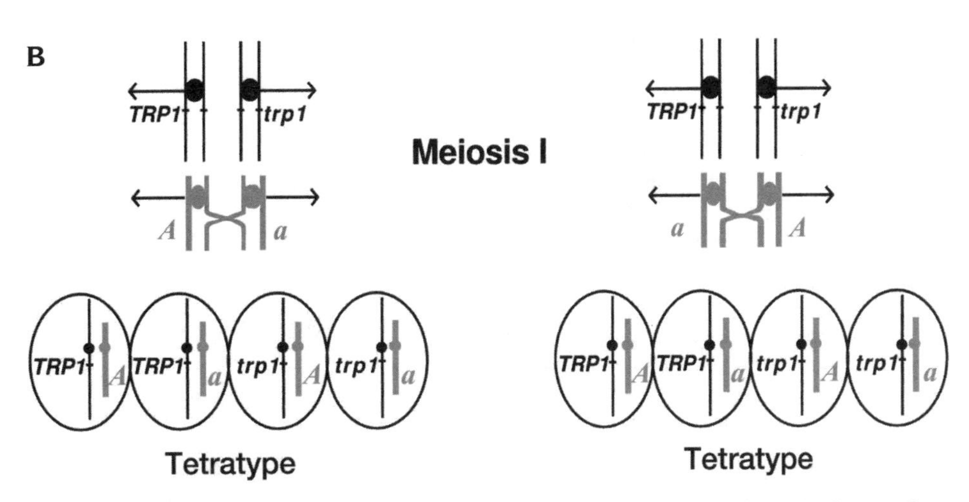

Figure 3. Using a known centromere-linked marker to examine centromere linkage of another gene. (*A*) In the cross *A, TRP1* x *a, trp1*, when no crossovers occur between our gene of interest (*A* in this example) and its centromere, gene *A* will show two types of segregation with respect to *TRP1*. The two outcomes are determined by the orientations with which the corresponding centromeres attach to the metaphase spindle. These meioses yield PD and NPD tetrads with equal likelihood. (*B*) Regardless of the orientations of the respective centromeres as they attach to the metaphase spindle, if there has been a single crossover between *A* and its centromere, what would have been a PD or NPD tetrad is converted into a T tetrad.

a single crossover between the gene of interest and its centromere and (2) the PD and NPD tetrads experienced no crossovers in this interval. Therefore, this mapping function severely underestimates the true map distance for intervals that experience multiple crossovers.

It is also possible to determine the percentage of SDS with reasonable accuracy if the hybrid contains two or more centromere-linked markers that may not be as close as, for example, *trp1*. In this case, the SDS array is decided on by the best agreement among the centromere-linked markers.

A Key Points

1) Phenotypes caused by mutations in a single gene segregate 2:2.
2) For unlinked genes, parental ditype (PD) = nonparental ditype (NPD).
3) For unlinked genes, at least one of which is not linked to a centromere, PD:NPD:T = 1:1:4.
4) For unlinked genes that are both linked to a centromere, tetratypes (T) result from crossing-over between the genes and the centromeres.
5) For linked genes, PD > NPD.
6) The map distance of small genetic intervals that experience at most two crossovers per meiosis can be estimated as

$$\text{centimorgans (cM)} = 100 \times \frac{1/2\ \text{T} + 3\ \text{NPD}}{\text{Total}}$$

B Decision Tree

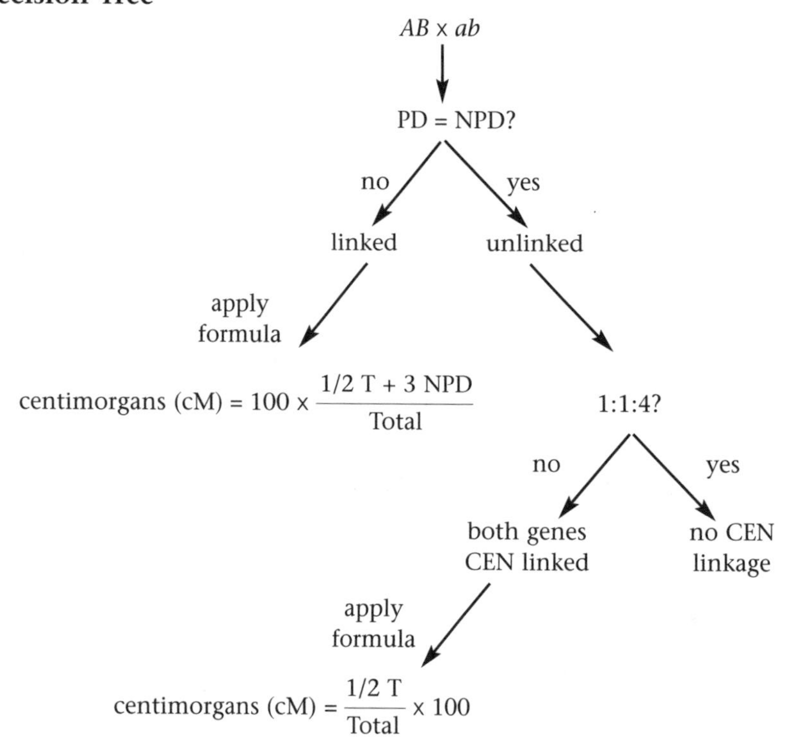

Figure 4. A tetrad analysis summary. (*A*) Key points for interpreting tetrad data. (*B*) A decision tree to aid in evaluating tetrad data.

Some of the key points to remember when analyzing tetrad data, and a decision tree to aid in drawing conclusions about genetic map relationships of the genes being analyzed, are shown in Figure 4.

AN INTRODUCTION TO TETRAD DISSECTION

Meiosis and sporulation are usually complete within four days following introduction of diploid cells to sporulation media. A sporulated culture contains a mixture of unsporulated diploid cells, asci with four haploid spores, and asci with fewer than four spores. The spores will not germinate or divide on sporulation medium. Most sporulated cultures may be stored in the refrigerator for several months with only a gradual loss of viability. The first step in dissection consists of treating the sporulated culture with enzymes that digest the ascus wall but do not disturb the association of the four spores from the ascus. Digested asci are carefully transferred onto a YPD plate using a sterile inoculating loop. The four spores are then separated from one another and moved to isolated positions on the YPD plate using a microneedle attached to a micromanipulator. The spores will germinate and form colonies 2–3 days after dissection. The spore clones can be transferred to slants or YPD plates for storage and analysis. A master plate is prepared by inoculating a YPD plate with the strains; most of the phenotypes are scored by replica-plating onto appropriate media. The mating types, complementarity of markers with similar phenotypes, and the nature of mutant alleles within a single gene can be determined by mating the spore clones to lawns of *MAT***a** and *MATα* tester strains with the appropriate genotypes. For example, we will score the *MAT***a** and *MATα* alleles of the haploid segregants of strain 3-3 by a test that scores their ability to mate with the tester strains 3-4 or 3-5 to form prototrophic diploids. The formation of the prototroph reflects both the mating type of the cells and the genetic complementation of the marker (*his1*) in the tester strains. Because we know the exact genotypes of the test lawns, the phenotypes of the diploids formed between the spore colonies and the various lawns allow us to deduce the genotypes of the spore colonies. A description of a variety of methods and instruments used for tetrad analysis can be found in Sherman and Hicks (1991).

STRAINS

3-1	GRY2501	*MAT***a**, *ura3-52, leu2-3-112, arg4-Δ42, trp2, cyh2*
3-2	GRY2502	*MATα, ura3-52, trp1-289, arg4-ΔBgl*II, *ade1*
3-3	3-1 x 3-2	
3-4	AAY1018	*MAT***a**, *his1*
3-5	AAY1017	*MATα, his1*
3-6	GRY2506	*MAT***a**, *arg4-Δ42, his3Δ1, trp1-289*
3-7	GRY2507	*MATα, arg4-Δ42, his3Δ1, trp1-289*
3-8	GRY2508	*MAT***a**, *arg4-ΔBgl*II, *his3Δ1, trp1-289*
3-9	GRY2509	*MATα, arg4-ΔBgl*II, *his3Δ1, trp1-289*
3-10	GRY2510	*MAT***a**, *trp1-289, ura3-52*
3-11	GRY2511	*MATα, trp1-289, ura3-52*
3-12	GRY2512	*MAT***a**, *trp2, ura3-52*
3-13	GRY2513	*MATα, trp2, ura3-52*

Note: Strain 3-3 may sporulate after long-term storage on YPD. The diploid may be reconstructed by mating strains 3-1 and 3-2.

PROCEDURE

Day 1

Preparation of the microneedle. Tetrad dissection needles can be made by hand pulling fine threads of glass or by using strands of fiber-optic glass as described in Techniques and Protocols #23, Making a Tetrad Dissection Needle. Preparation of a needle will be demonstrated and supplies provided so that you can make your own. Needles and holders are also commercially available (for suppliers, see http://www.biosupplynet.com).

Treatment with Zymolase. Your instructors introduced the diploid strain 3-3 to sporulation medium three days before the beginning of the course. Examine the sporulated culture under a microscope and identify the unsporulated cells, the four-spored asci, and the asci with fewer than four spores. Each student will be allotted three days to produce 20 dissected tetrads with four viable spores. The methods used to treat cells with Zymolase, prepare a dissection plate, and micromanipulate tetrads are described in Techniques and Protocols #22, Tetrad Dissection.

Day 3

Complete dissection of tetrads.

Day 5

Using sterile flat toothpicks and following the grid provided (see Appendix D, Templates for Making Streak Plates), each student must prepare two master plates with spore colonies that have four viable members (ten tetrads per plate). Include the two parental control strains on each plate. Incubate overnight at 30°C.

Day 6

Replica-plate tetrads to test media. Replica-plate the master plate to these media: YPD, SC-leu, SC-arg, SC-ade, SC-trp, YPD+cyh, and six more YPD plates (see hints below before you begin). The first YPD plate serves as a fresh master plate for future use. The remaining YPD plates are for matings of your spore colonies with the tester strains. Incubate all plates at 30°C.

Hints: Twelve replicas is a large number to make from one master plate. After the first six replicas, place a fresh velvet on the replica-plating block, and make a fresh print from your master plate. Because you are using the master plate twice, be careful to make a light print on the first velvet so that there will be cells remaining on the master plate for the second print. Examine each replica-plate after you produce it. You should be able to see a light deposition of cells corresponding to the pattern on your master plate. If you do not, try again. (As an alternative method, sterile H_2O in a microtiter dish can be inoculated with spore colonies and the cells can be transferred to the different test plates using the multiprong inoculator [frogging technique]. This will be demonstrated by the instructors.) There can be cross-feeding between *trp1* and *trp2* strains, so it is important that the print to the SC-trp plate be light.

Preparation of mating-type and allele testers. Inoculate 5 ml of YPD with the tester strains 3-4 through 3-13. Alternatively, make large patches of tester strains on YPD plates. Incubate overnight at 30°C.

Day 7

Preparation of mating-type test lawns. Use either method.

a. Centrifuge the 5-ml cultures of 3-4 and 3-5 (5 min at 2000 rpm in a tabletop centrifuge). Resuspend in 4 ml of YPD in a sterile screw-top tube. Add 0.15 ml of sterile H_2O and 0.15 ml of the resuspended 3-4 culture to two SD plates (store the remaining cells at 4°C). Distribute the inoculum evenly with a sterile rod, place the plate on a level surface, and allow the plate to dry. Repeat with the 3-5 cells.

b. Use an applicator stick to scrape 3-4 cells from a fresh patch on a YPD plate and resuspend in 4 ml of YPD liquid. These cells should be more concentrated than a saturated overnight culture. Spread 0.2 ml on two SD plates. Repeat process for 3-5 cells. Let the liquid dry in before you replica-plate onto the lawns.

Preparation of allele and complementation test lawns. These lawns are prepared using the same methods used for the mating-type lawns except that both the *MAT***a** and *MATα* versions of the test strain are combined on the same plate. Prepare four types of test lawns and make two of each type. Spread the *arg4* allele test lawns (3-6 mixed with 3-7, and 3-8 mixed with 3-9) on SC-ura-his. For the Trp complementation lawns (3-10 mixed with 3-11, and 3-12 mixed with 3-13), spread the cells onto SC-trp plates. Let the liquid dry in before you replica-plate onto the lawns.

Replica-plating to test lawns. Use the six YPD replicas of your tetrads that you made yesterday as the source of the cells for printing onto your six test lawns. For each test lawn, place a fresh velvet on the replica-plating block, make a print on the velvet with one of your YPD plates, and place the test lawn plate on the velvet to transfer the tetrads onto the test lawn. Incubate the test plates overnight at 30°C.

Day 9

Replica-plate the SC-ura-his plates with the *arg4* allele-testing lawns to SC-arg plates and refrigerate. At the end of the day, these plates will be collected by your laboratory assistant, who will irradiate them with 7500 μJ of UV light (254 nm) using a Stratagene Stratalinker (set on 75).

Score the growth on the mating-type test plates and the Trp complementation plates. Because *trp1* defective strains can cross-feed *trp2* defective strains, it may be necessary to replica-plate the SC-trp plate to a fresh SC-trp plate and score that second plate tomorrow.

Day 10/11

Score the *arg4* allele-testing plates and the Trp complementation plates.

Day 11

Determine the number of PD, NPD, and T tetrads for each pairwise combination of markers segregating in your cross. Record the PD/NPD/T data on the scoring sheet at the end of this experiment. In addition, record the frequency of second-division segregation for each marker. Determine the distance between each gene and its centromere by searching for map information about each gene on the *Saccharomyces* Genome Database (SGD; http://www.pathway.yeastgenome.org/). Use the data you have gathered to calculate

1. The map distance for each gene scored from all other genes scored
2. The map distance between each gene and its centromere (how does this compare to the information provided by the SGD?)
3. The gene conversion frequencies (3:1 and 1:3 segregation) for all alleles that we studied

Day 14

Be prepared to present your data.

MATERIALS

Note: Amount per student: 20 tetrads. Each group requires double these quantities.

Day 1

Capillary pipettes for preparing microneedle holders
Super Glue
Fiber-optic glass
Sporulated culture of strain 3-3
Zymolase 100T (120493-1, Seikagaku America Inc.) solution (0.5 mg/ml in 1 M sorbitol)
Sterile distilled H_2O
4 YPD plates

Day 5

2 YPD plates
Sterile toothpicks

Day 6

2 plates each of
- SC-ade
- SC-arg
- SC-leu
- SC-trp
- YPD+cyh

14 YPD plates
4 sterile velvets
10 tubes containing 5 ml of YPD liquid or 5 YDP plates

Day 7

10 tubes containing 4 ml of YPD
Sterile distilled H_2O
4 SD plates
4 SC-his-ura plates
4 SC-trp plates
12 sterile velvets

Day 9

4 SC-arg plates
4 SC-trp plates
8 sterile velvets

REFERENCES

Fincham J.R.S., Day P.R., and Radford A. 1979. *Fungal genetics*. University of California Press, Berkeley.

Mortimer R.K. and Hawthorne D.C. 1966. Genetic mapping in *Saccharomyces*. *Genetics* **53:** 165–173.

——. 1969. Yeast genetics. In *The yeasts* (ed. A.H. Rose and J.S. Harrison), vol. 1, pp. 385–460. Academic Press, New York.

Mortimer R.K. and Schild D. 1981. Genetic mapping in *Saccharomyces cerevisiae*. In *The molecular biology of the yeast* Saccharomyces: *Life cycle and inheritance* (ed. J.N. Strathern et al.), pp. 11–26. Cold Spring Harbor Laboratory, Cold Spring Harbor, New York.

Perkins D.D. 1949. Biochemical mutants in the smut fungus *Ustilago maydis*. *Genetics* **34:** 607–626.

Sherman F. and Hicks J. 1991. Micromanipulation and dissection of asci. *Methods Enzymol.* **194:** 21–37.

Experiment III: Tetrad analysis; 3-3 Group no.______

Names______________________________________

	MAT	TRP1	TRP2	CYH2	ARG4	LEU2	ADE1
P MAT **N** **T**							
P TRP1 **N** **T**							
P TRP2 **N** **T**							
P CYH2 **N** **T**							
p ARG4 **N** **T**							
P LEU2 **N** **T**							

%SDS (with TRP1)							
Gene/ CEN Distance (SGD)							

EXPERIMENT IV

Mitotic Recombination and Random Spore Analysis

In diploid strains, alterations leading to loss of genetic information can be generated by mitotic crossing-over, gene conversion, or chromosome loss. Because such events are usually observed as the loss of one of two alleles in a heterozygous strain, they are referred to as loss of heterozygosity (LOH) events. Mitotic crossing-over and gene conversion are illustrated in Figure 1. Mitotic crossing over results in LOH of all markers located distal to the point of exchange on the chromosome arm. Mitotic gene conversion results in nonreciprocal exchange of only a small chromosomal region (i.e., a single gene or closely linked genes) and is believed to be analogous to the irregular segregations that are observed at low frequency after meiosis (meiotic gene conversion). Chromosome loss, which results in LOH of markers on both arms of a chromosome, produces a 2N-1 diploid that may or may not have impaired growth, depending on the particular chromosome.

Mitotic crossing-over between two nonsister chromatids is likely to occur during the G_2 stage of the cell cycle (see Fig. 1). After mitosis, a crossover event can lead to two daughter cells that are homozygous for part of the chromosome arm that was heterozygous in the mother cell (this will occur in half of the recombination events because of the random orientation of chromosome pairs on the mitotic spindle). Thus, mitotic recombination can lead to the "uncovering" of recessive markers. For some markers, "papillae" or "sectors" derived from the homozygous daughter cells can be observed in a background of heterozygous cells, resulting in papillating or sectoring colonies. As a result of mitotic crossing-over and subsequent mitotic segregation, all markers distal to the site of recombination on the chromosome arm are simultaneously homozygosed. It is possible to generate a recombinational map of the chromosomal arm using mitotic recombination data. An approximately linear relationship exists between the frequency of homozygosis of a gene and its distance from the centromere. Thus, quantitation of the relative frequencies of different classes of recombinants can be used to determine the relative distances between two markers and between a marker and its centromere.

Mitotic gene conversion also results from recombination between two nonsister chromatids in G_2 (see Fig. 1). In this case, only a limited chromosomal region is

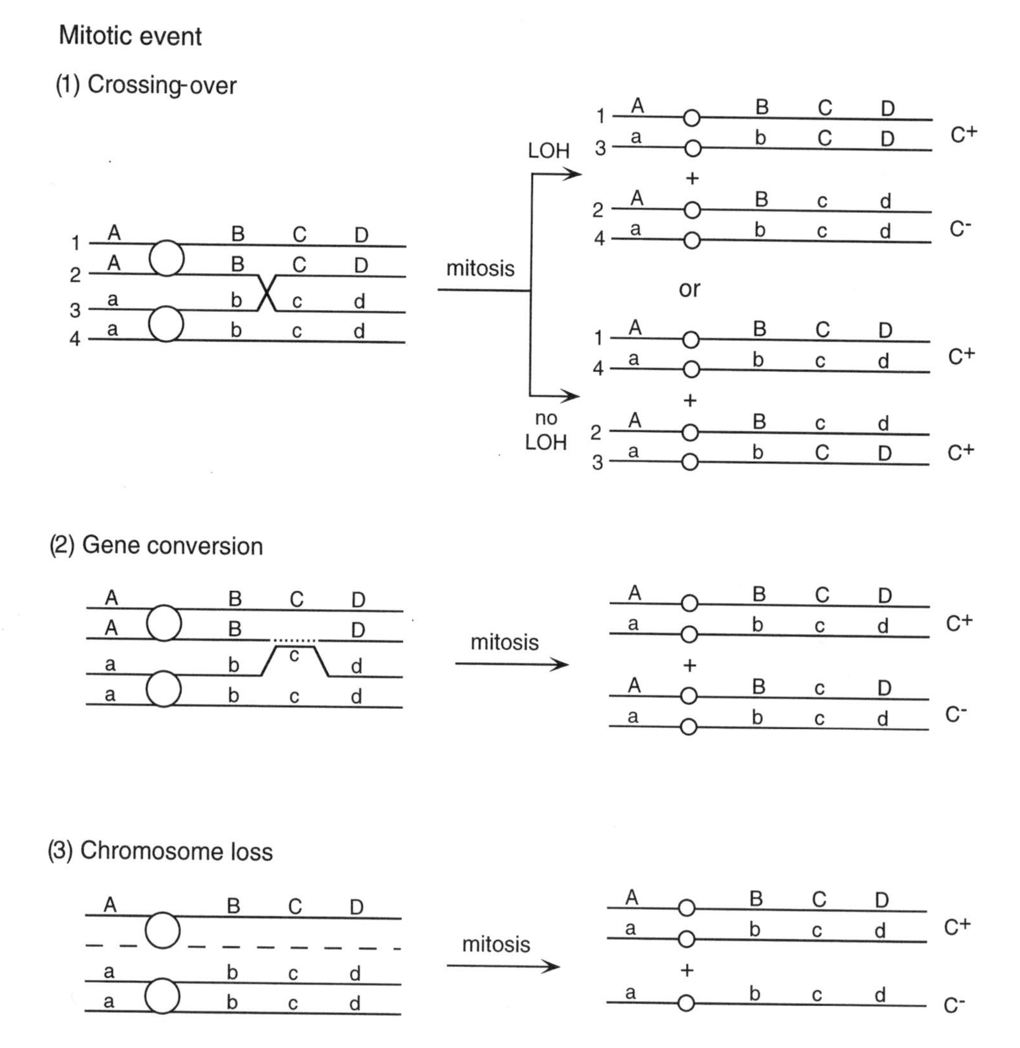

Figure 1. Mitotic segregation of heterozygous markers in a diploid strain after either (1) crossing-over, (2) mitotic gene conversion, or (3) chromosome loss.

homozygosed, with the result that distal markers retain their heterozygosity. If two or more markers sector together, this event is most likely due to mitotic crossing-over and not to gene conversion, because conversion tracts are generally short (several kilobases at most).

It is important to note that the spontaneous rate of mitotic recombination is low, ranging from approximately 10^{-6} to 10^{-4}, depending on the distance of a particular gene from its centromere and other less well-defined factors. Chromosome loss events occur at similarly low frequencies. To find rare LOH events easily, it is usually necessary to apply a selection or screen. In this experiment, we will use selections for canavanine resistance and 5-fluoro-orotic acid (5-FOA) resistance to identify cells in which a LOH event has occurred.

EXPERIMENTAL DESIGN

A diploid will be created by mating the following two haploid strains:

4-1 TSY812 *MATα can1 hom3 leu2 lys2 ura3*
4-2 TSY813 *MAT***a** *ade2 his1 lys2 trp1*

URA3 and *CAN1* are on the left arm of chromosome V, and *HIS1* and *HOM3* are on the right arm of chromosome V. The *URA3* and *CAN1* genes are particularly useful for this experiment because they have the unusual feature that strong selection techniques exist for recessive alleles of both genes. Starting with a diploid that is heterozygous at both loci, *ura3* strains that arise by mitotic recombination, chromosome loss, or meiosis can be selected by their resistance to 5-FOA, and *can1* strains by their resistance to canavanine.

The *URA3* gene encodes orotidine-5´-phosphate decarboxylase, an enzyme that catalyzes one of the steps in pyrimidine synthesis. 5-FOA is taken up by cells and is converted to the toxic compound 5-fluorouracil by the action of the decarboxylase. *URA3/URA3* homozygotes, *URA3/ura3* heterozygotes, and *URA3* hemizygotes (one copy of a gene in a diploid strain) are all able to convert 5-FOA to 5-fluorouracil, and are sensitive to 5-FOA (5-FOA^S), whereas *ura3/ura3* homozygotes and *ura3* hemizygotes lack the decarboxylase activity and are resistant to 5-FOA (5-FOA^R).

The *CAN1* gene encodes the arginine permease, which allows uptake of arginine from the medium. Canavanine is a toxic analog of arginine that is taken up by cells through the arginine permease. *CAN1/CAN1* homozygotes, *CAN1/can1* heterozygotes, and *CAN1* hemizygotes are all able to take up canavanine, and so are sensitive to canavanine (Can^S), whereas *can1/can1* homozygotes and *can1* hemizygotes lack the permease and are resistant to canavanine (Can^R).

We will find the frequency of LOH at each locus by determining the frequency of Can^R and 5-FOA^R mitotic segregants that arise from the diploid strain 4-1 x 4-2. We will compare these frequencies to the frequency of spontaneous mutations in *CAN1* and *URA3* in the haploid strain 4-2. Mitotic recombination will be distinguished from chromosome loss by examining the Hom3 phenotype on the opposite arm of chromosome V. We will determine the relative positions of *CAN1* and *URA3* with respect to the centromere on chromosome V by seeing how often LOH occurs at *CAN1* but not at *URA3*, or at *URA3* but not at *CAN1*.

An additional part of this experiment will be to examine meiotic recombination in the 4-1 x 4-2 diploid using random spore analysis. This method allows meiotic products to be isolated and scored without performing tetrad analysis. It is particularly useful when many crosses must be analyzed or a rare recombinant must be identified. It is important to realize, however, that information about centromere linkage, potential aneuploidy, and the ability to use the tetrad mapping function for more accurate determination of genetic distance is lost when random spore analysis is used in place of tetrad analysis. The principle behind the method is that haploid meiotic segregants are selected away from unsporulated diploid cells by selecting for canavanine resistance in a strain that was originally a *CAN1/can1* heterozygote. Half of all meiotic segregants will be Can^R because they will have received the *can1* allele.

STRAINS

4-1	TSY812	*MATα can1 hom3 leu2 lys2 ura3*
4-2	TSY813	*MAT***a** *ade2 his1 lys2 trp1*
4-3	GRY2426	*MAT***a** *his3Δ1 met15Δ0*
4-4	GRY2427	*MATα his3Δ1 met15Δ0*

PROCEDURE

Day 1

Subclone strains 4-1 and 4-2 to a YPD plate. Incubate at 30°C.

Day 3

In the morning, mate strains 4-1 and 4-2. Transfer a single colony of 4-1 and a single colony of 4-2 into a sterile microfuge tube. Add 200 µl of H_2O, vortex, and pellet for 30 seconds. Draw off most of the liquid and transfer the cells to a YPD plate. Incubate for 4 hours at 30°C. Streak mating cells across the top of the YPD plate. Use a tetrad dissecting microscope and micromanipulator to identify dumbbell-shaped zygotes and move them to an open area of the YPD plate. Incubate at 30°C.

In addition, streak out cells from the mixed patch to an SD+lys plate to select for diploids. Incubate at 30°C.

Day 5

Patch zygotes to YPD and SD+lys to confirm that they are correct. Include parents as controls. Carefully pick diploids from the SD+lys plate and patch onto YPD and SD+lys plates.

Day 6

Check the SD+lys plates to identify 4-1 x 4-2 diploids.

LOH. In the evening, inoculate 5 ml of YPD with a 4-1 x 4-2 diploid and 5 ml of YPD with the 4-2 haploid as a control. Incubate on the roller drum at 30°C.

Meiotic recombination. To prepare the 4-1 x 4-2 diploid for sporulation, patch onto a YPD plate and incubate overnight at 30°C.

Day 7

Mutation and LOH. Prepare tenfold serial dilutions of the saturated culture of the 4-1 x 4-2 diploid in sterile microfuge tubes. To determine the frequency of LOH, plate 100 µl of 10^{-1}, 10^{-2}, and 10^{-3} dilutions on SC+5-FOA and SC-arg+canavanine plates. To determine the number of viable cells, plate 100 µl of 10^{-4} and 10^{-5} dilutions on YPD plates. To determine the frequency of spontaneous mutation at the *CAN1* and *URA3* genes, prepare similar dilutions of the 4-2 haploid culture. For these cells, plate 100 µl of 10^{-1} and 10^{-2} dilutions on SC+5-FOA and SC-arg+canavanine plates. To determine the number of viable cells, plate 100 µl of 10^{-4} and 10^{-5} dilutions on YPD plates. Incubate all plates at 30°C.

Meiotic recombination. Transfer the 4-1 x 4-2 diploid from the YPD plate to liquid sporulation medium and incubate at 25°C according to Techniques and Protocols #10, Random Spore Analysis.

Day 10

Mutation and LOH. Count the colonies on the selective and YPD plates and calculate the frequencies of Can^R and 5-FOA^R for the 4-1 x 4-2 diploid and the 4-2 haploid. With the multipronged inoculating device (frogger), spot 22 independent isolates of each type of the drug-resistant strains onto SC+5-FOA, SC-arg+canavanine, SC-met (to score *HOM3*), and YPD plates. If you do not have enough 5-FOA^R or Can^R isolates of strain 4-2, just plate what you have. Include the 4-1 x 4-2 diploid and the parental haploids 4-1 and 4-2 as controls.

Meiotic recombination. Check the sporulating culture of 4-1 x 4-2. If the culture has sporulated, digest and break apart tetrads, and plate dilutions of the random spores according to Techniques and Protocols #10, Random Spore Analysis.

Day 11

Mutation and LOH. On the basis of growth of patched or spotted strains, calculate the frequency of Can^R for strains first selected on 5-FOA and 5-FOA^R for strains first selected on canavanine.

Day 13

Using the multipronged device, patch or spot 45 independent random spore colonies from the 4-1 x 4-2 diploid onto three YPD plates, and one each of SC-ura, SC-his, SC-met, and SC-arg+canavanine plates. Include the 4-1 x 4-2 diploid and the parental haploids 4-1 and 4-2 as controls.

Start mating-type tester strains 4-3 and 4-4 on YPD.

Day 14

Test mating type of random spore clones. Replica a plate from the YPD plate of frogged cultures to a lawn of 4-3 cells in 0.2 ml of YPD spread on an SD plate. Repeat with a lawn of 4-4 cells on an SD plate.

Day 15

Score the mating-type plates and discount strains that do not mate well, because they might not be haploid segregants. In addition, be suspicious of isolates that show a mix of red (*ade2*) and white (*ADE2*) cells on the YPD plate.

Day 16

Meiotic recombination. Based on the growth of spotted spore colonies, determine the frequency of meiotic recombination among the *CAN1, URA3, HIS1,* and *HOM3* genes.

MATERIALS (per student)

Day 1

1 YPD plate

Day 3

1 YPD plate
1 SD+lys plate
Sterile loop or toothpicks
Sterile H_2O

Day 5

1 YPD plate
1 SD+lys plate

Day 6

1 YPD plate
2 culture tubes containing 5 ml of YPD

Day 7

Sterile microcentrifuge tubes
5 SC-arg+canavanine plates
5 SC+5-FOA plates
4 YPD plates
Liquid sporulation medium
Sterile H_2O
Glass spreader
70% ethanol for sterilization

Day 10

5 SC-arg+canavanine plates
2 SC+5-FOA plates
2 SC-met plates
2 SC-leu plates
2 SC-his plates
2 YPD plates
Sterile toothpicks/dowels
Multipronged inoculating device
2 sterile 96-well plates
Materials for breaking apart tetrads from Techniques and Protocols #10, Random Spore Analysis

Day 13

1 SC-arg+canavanine plate
1 SC-ura plate
1 SC-met plate
1 SC-his plate
4 YPD plates
Sterile toothpicks/dowels
Multipronged inoculating device
1 sterile 96-well plate

Day 14

2 SD plates
Sterile velvets

DATA

Group #

Spontaneous mutation (strain 4-2)

Medium	Dilution	# of Colonies	Freq. of Mutation
5-FOA	10^{-1}		
	10^{-2}		
CAN	10^{-1}		
	10^{-2}		
YPD	10^{-4}		
	10^{-5}		

Loss of heterozygosity (strain 4-1 x 4-2)

Medium	Dilution	# of Colonies	Freq. of LOH
5-FOA	10^{-1}		
	10^{-2}		
	10^{-3}		
CAN	10^{-1}		
	10^{-2}		
	10^{-3}		
YPD	10^{-4}		
	10^{-5}		

5-FOA^{R} (strain 4-1 x 4-2)

#	Can	Met
1		
2		
3		
4		
5		
6		
7		
8		
9		
10		
11		
12		
13		
14		
15		
16		
17		
18		
19		
20		
21		
22		
4-1		
4-2		
4-1 x 4-2		

Can^{R} (strain 4-1 x 4-2)

#	5-FOA	Met
1		
2		
3		
4		
5		
6		
7		
8		
9		
10		
11		
12		
13		
14		
15		
16		
17		
18		
19		
20		
21		
22		
4-1		
4-2		
4-1 x 4-2		

Meiotic recombination (random spore analysis)

#	Can^R	Ura	His	Met
1				
2				
3				
4				
5				
6				
7				
8				
9				
10				
11				
12				
13				
14				
15				
16				
17				
18				
19				
20				
21				
22				
4-1				
4-2				
4-1 x 4-2				

#	Can^R	Ura	His	Met
23				
24				
25				
26				
27				
28				
29				
30				
31				
32				
33				
34				
35				
36				
37				
38				
39				
40				
41				
42				
43				
44				
45				

EXPERIMENT V

Transformation of Yeast

Saccharomyces cerevisiae is unique among eukaryotes because of the ease with which it can be transformed with DNA, and the high frequency with which the introduced DNA undergoes homologous recombination with genomic DNA. There are several requirements for a successful transformation experiment: (1) a means of introducing DNA into cells, (2) a selectable marker on the introduced DNA with corresponding nonreverting alleles in the chromosome, and (3) vector systems that allow propagation of cloned DNA in both *Escherichia coli* and yeast.

TRANSFORMATION METHODS

The original method for yeast transformation involved incubating spheroplasted cells with DNA in the presence of polyethylene glycol (PEG) and $CaCl_2$ (Hinnen et al. 1978). A more convenient and much more widely used method involves the treatment of cells with the alkali salt, lithium acetate (LiAc), followed by incubation with DNA and PEG (Ito et al. 1983). DNA can also be introduced by electroporation (Hashimoto et al. 1985; Becker and Guarente 1991), whereby a brief electrical pulse permeabilizes the cells to DNA; by agitation of cells with glass beads (Costanza and Fox 1988); by bombardment of cells with DNA-coated particles (currently the only way to transform mitochondria) (Fox et al. 1988; Johnston et al. 1988); and by direct conjugation between bacterial and yeast cells (Heinemann and Sprague 1989). The method of choice depends on the purpose of the experiment, the number of strains to be transformed, and the desired number of transformants. The efficiency of transformation is often the most important parameter: If the goal is simply to put a plasmid into a given strain (only a few colonies needed), then any of the methods will work with almost any strain of yeast; if the goal is to get 10^6 transformants for screening a library, the strain and transformation method must be carefully chosen. Spheroplasting, LiAc, and electroporation can all give high transformation frequency under optimal conditions.

SELECTABLE MARKERS

Although the frequency of transformation of yeast can be quite high, only a small fraction of the total number of cells in a transformation experiment becomes transformed.

Therefore, it is essential to have reliable selectable markers to select for those cells that have become transformed. The most common selectable markers used in yeast transformation are those that complement a specific auxotrophy. For example, the yeast *LEU2* gene encodes β-isopropylmalate dehydrogenase and complements the leucine auxotrophy of a *leu2* mutant. Other commonly used selectable markers are *URA3*, mutations of which result in uracil auxotrophy; *HIS3*, mutations of which result in histidine auxotrophy; and *TRP1*, mutations of which result in tryptophan auxotrophy. Of equal importance to the selectable marker is the corresponding chromosomal mutation that causes the auxotrophy. This mutation should be completely recessive and nonreverting; for example, the *leu2-3,112* mutation is a double frameshift mutation that reverts with a very low frequency ($<10^{-10}$) and is completely complemented by the wild-type *LEU2* gene. More recently, dominant drug resistance has been used as a selectable marker in yeast, because it is used in bacterial transformation (Hadfield et al. 1990). This has the potential advantage that the selectable marker lacks any homology with the yeast genome.

VECTOR SYSTEMS

Yeast cells that have taken up DNA during the transformation process can maintain that DNA, and thus become transformed, either by integration of the DNA into a chromosome or by autonomous replication. Integration into a chromosome takes place almost exclusively by homologous recombination in yeast. Once integrated, the transforming DNA is part of the chromosome and segregates in mitosis and meiosis with the same high fidelity as a chromosome. Plasmids used for integration have a yeast selectable marker, but no other yeast elements. Autonomous replication requires that the transforming DNA have a yeast origin of DNA replication. Originally called ARS (autonomously replicating sequence) elements, these can have either chromosomal DNA replication origins or originate from the endogenous yeast 2μ plasmid. Because the yeast replication origin is a relatively simple and short DNA sequence, DNA from other organisms will occasionally be found to have yeast ARS activity. To transform yeast cells stably, transforming DNA must either have sufficient homology to the yeast genome to integrate, or carry an ARS element.

Plasmids with only a chromosomal ARS element (ARS plasmid) have a variable copy number and often fail to segregate to the daughter cell in a division (Murray and Szostak 1983), resulting in a high rate of plasmid loss. Autonomously replicating plasmids may also have a centromere, or CEN element. A CEN/ARS plasmid (CEN plasmid) is more stable than a simple ARS plasmid because the centromere mediates the attachment of the plasmid to the mitotic spindle, ensuring segregation to both mother and daughter cells. Because of the high fidelity of segregation, the copy number is maintained at one or two plasmids per cell. CEN plasmids typically show a 2:2 (if the cell had one copy) or 4:0 (if the cell had at least two copies) segregation pattern in meiosis.

An autonomously replicating plasmid that has the 2μ origin of replication (2μ-based plasmid) segregates in mitosis with about the same fidelity as a CEN plasmid, but is present at a much higher copy number, typically 20–50 copies per cell. The high fidelity of segregation of 2μ plasmids depends on the presence of the endogenous 2μ plasmid. If a strain lacks the endogenous 2μ plasmid, then introduced 2μ-based plasmids will segregate as ARS plasmids (Murray and Szostak 1983).

Many older yeast plasmids were named systematically. In this system, integrating plasmids were designated YIp; ARS plasmids, YRp; CEN plasmids, YCp; and 2μ plasmids, YEp (E for episome). YRp plasmids are rarely used because of their extreme instability.

In addition to the yeast-specific elements, all standard yeast vectors also have a bacterial origin of replication and bacterial selectable marker, usually ampicillin resistance. Older vectors (YIp5, YEp24, YCp50) are usually based on a pBR322 backbone. More recent plasmids are often based on plasmid backbones that replicate to a higher copy number in bacteria, and have useful polylinker sequences and single-stranded phage origins for the isolation of DNA for sequencing. A well-designed set of yeast vectors was described by Sikorski and Hieter (1989) and is used in this experiment. Figure 1 illustrates the plasmids to be used, all based on pRS306, an integrating vector with *URA3* as the selectable marker: pRS316 is pRS306 with an inserted CEN/ARS element, and pRS426 is a derivative of pRS306 with an inserted 2μ origin (the orientation of the polylinker is also reversed). pJS801 is a derivative of pRS306 with a genomic fragment including the yeast *LEU2* gene inserted in the polylinker.

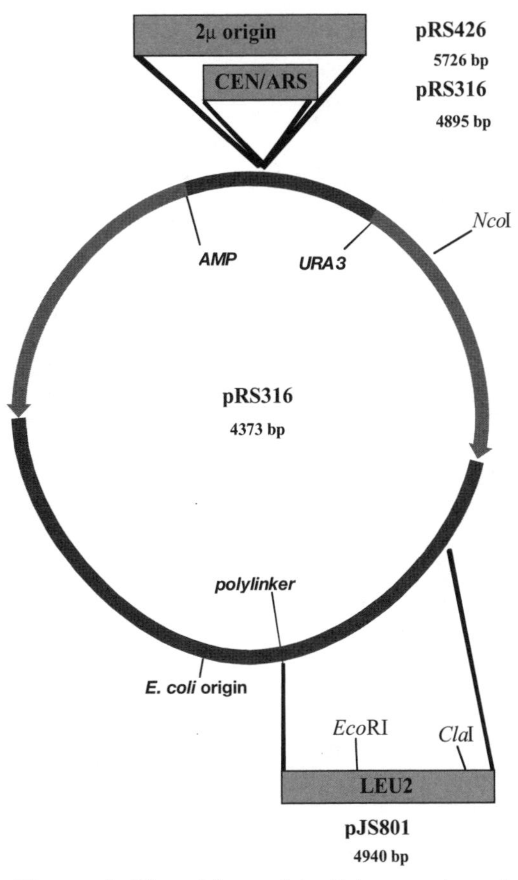

Figure 1. Plasmids used in this experiment.

EXPERIMENT V(A)

INTEGRATION

In most cases, integration of a circular plasmid into the yeast genome occurs by a single crossover and yields a direct repeat of the yeast sequence on the plasmid, as shown in Figure 2. Note that the entire plasmid is integrated, including the bacterial sequences and the yeast selectable marker—these now serve as a physical and phenotypic marker for the site of integration. The plasmid used in this experiment is pJS801, which has two regions of homology with the yeast genome: the *URA3* gene that is part

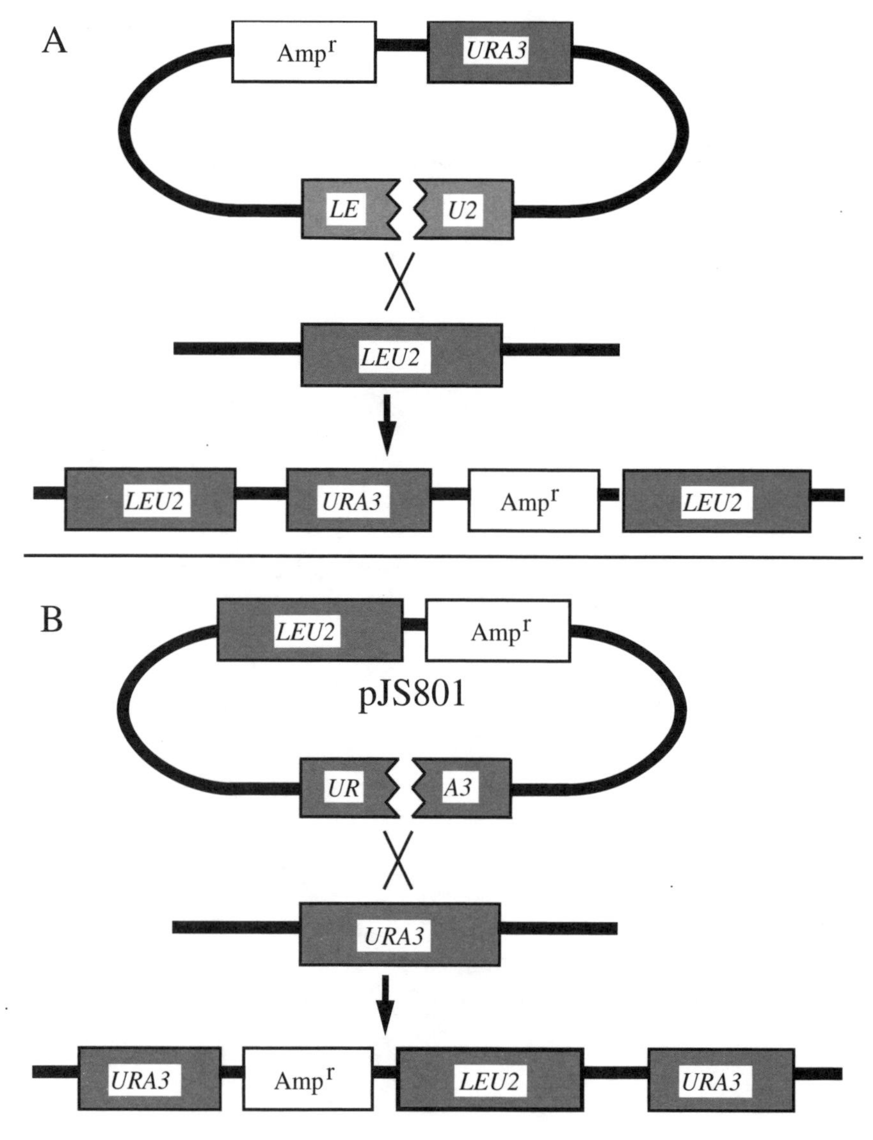

Figure 2. Integration of plasmid pJS801 at either the (*A*) *LEU2* or (*B*) *URA3* locus.

of the parent vector, pRS306, and the *LEU2* gene that was inserted into the polylinker of pRS306.

pJS801 cannot replicate in yeast because it lacks an ARS element and can, therefore, transform yeast only by integration. Integration can occur at either the *URA3* locus (actually the *ura3-52* locus in the strain used here) or the *LEU2* (*leu2-3,112*) locus by homologous recombination (Fig. 2). Cutting a plasmid within a region of homology with the genome greatly increases the specificity and frequency of recombination at that site (Orr-Weaver et al. 1981); the free ends are highly recombinogenic and increase the efficiency of integration, and thus transformation, by approximately 100-fold. In this experiment, uncut pJS801 should transform strain 5-1 with a very low efficiency and integrate at either *ura3* or *leu2*. Cutting pJS801 at the unique *Nco*I site in *URA3* should result in nearly 100% integration at *ura3*, and cutting pJS801 at the unique *Eco*RI site in *LEU2* should result in nearly 100% integration at *leu2*. Note that some of the transformants with pJS801 integrated at *leu2* will be Leu⁻ because of gene conversion associated with the double-strand-break repair.

Integration at a locus can be verified both genetically and physically. In this experiment, transformants will be crossed to tester strains to determine the site of integration. Integration at *leu2* will be tested by crossing to a strain with the *LEU2* allele. If the plasmid integrates at the *LEU2* locus, then the plasmid-borne *URA3* gene will be genetically linked to the *LEU2* locus; thus, every tetrad should be a parental ditype (PD) with two Ura⁺ Leu⁺ and two Ura⁻ Leu⁺ spores. Integration at *ura3* will be tested by crossing to a *URA3* strain. If the plasmid integrates at the *ura3* locus, then the plasmid-borne *URA3* gene will be genetically linked to the *ura3* locus; thus, every tetrad should be a PD with four Ura⁺ spores. As an exercise, predict the outcome of the cross to a *URA3* strain if the plasmid had integrated at *leu2* instead of *ura3*.

EXPERIMENT V(B)

REPLICATION AND STABILITY

Strain 5-1 will also be transformed with the autonomously replicating plasmids pRS316 and pRS426. The frequency of transformation will be higher for these plasmids than for the integrating plasmid. The stability of these plasmids in mitosis and meiosis will be examined.

STRAINS

5-1	TSY623	*MATα ade2-101 his3-Δ200 leu2-3,112 ura3-52*
5-3	TSY1017	*MAT***a** *his3-Δ200 leu2-3,112 trp1-1 ura3-52*
5-4	TSY808	*MAT***a** *lys2-801*

PLASMIDS

pJS801 *LEU2 URA3*
pRS316 *CEN URA3*
pRS426 2μ *URA3*

PROCEDURE

Day 1

Subclone strain 5-1 on a YPD plate. Incubate at 30°C.

Day 3

In the morning, inoculate 5 ml of YPD with a single colony of strain 5-1 to make a preculture. Incubate at 30°C with agitation. In the evening, determine the cell concentration using a hemocytometer. Assuming an approximate generation time of 100 minutes, calculate the volume of preculture to be added to 100 ml of YPD for the culture to reach a concentration of 2–10^7 cells/ml by 10 AM on day 4.

Day 4

Harvest the cells by centrifugation at 2000 rpm in a clinical centrifuge and follow the protocol for LiAc transformation given in Techniques and Protocols #1, High-efficiency Transformation of Yeast. Transform 5-1 under the conditions listed below.

Experiment	Recipient	DNA	Digestion	Selective Medium
A	5-1	pJS801	none	SC-ura
	5-1	pJS801	*Nco*I	SC-ura
	5-1	pJS801	*Eco*RI	SC-ura
B	5-1	pRS316	none	SC-ura
	5-1	pRS426	none	SC-ura
Control	5-1	none	none	SC-ura

Day 6

1. Replica-plate the pJS801 transformation plates to SC-leu and SC-ura plates.

Day 7

1. To analyze the transformants genetically, we will cross them to tester strains. For the pJS801 transformants, carefully pick two Ura⁺ Leu⁺ colonies from each of the

transformation plates with sterile toothpicks or a loop and mix them with the tester strain 5-4 in a small patch on a YPD plate. The meiotic stability will be tested in diploids made by mating to tester strain 5-3. Mate one pJS801 *Nco*I transformant to tester strain 5-3. For the pRS316 and pRS426 transformants, carefully pick two Ura$^+$ colonies and mix with the tester strain 5-3 in a small patch on a YPD plate. Incubate mating mixes overnight at 30°C.

2. The assays for mitotic stability require that transformants be purified away from the untransformed cells on the transformation plates. Purify two transformants each from the *Nco*I-cut integrating transformation (Expt. V[A] on day 4) and the pRS316 and pRS426 plasmid transformations (Expt. V[B] on day 4) by streaking onto SC-ura agar. Incubate at 30°C.

Day 8

Replica-plate the YPD plates containing the mating mixes onto the appropriate selective medium described below. Only diploid cells formed by mating should be able to grow in the patches where the cells were mixed. Incubate at 30°C.

Transformant	Tester	Selective Medium
Integration location		
5-1+pJS801 (uncut)	5-4	SD
5-1+pJS801 (*Nco*I)	5-4	SD
5-1+pJS801 (*Eco*RI)	5-4	SD
Meiotic stability		
5-1+pJS801 (*Nco*I)	5-3	SD+his+leu
5-1+pRS316	5-3	SD+his+leu
5-1+pRS426	5-3	SD+his+leu

Day 9

1. In preparation for sporulation, patch the diploids growing on the selective plates to a YPD plate and incubate for 1 day at room temperature.
2. Inoculate one 5-ml YPD culture with a toothpickful of cells from one colony of each purified transformant from the SC-ura plates (total of six cultures). Place the tubes on the roller drum overnight at 30°C.

Day 10

1. Transfer the diploids from YPD to sporulation medium (both liquid and plate to compare efficiency of sporulation) and incubate several days at room temperature.
2. Make serial dilutions of the overnight 5-ml YPD cultures in sterile H_2O and spread aliquots on YPD plates. Try to get 100–300 colonies/plate. Incubate at 30°C.

Day 13

1. Begin dissecting tetrads from the crosses involving the transformants. Try to dissect ten tetrads from each cross. Incubate tetrad plates at 30°C.
2. Replica-plate YPD plates with the appropriate number of colonies to SC plates and SC-ura plates. Incubate at 30°C.

Days 15–16

1. As tetrads grow up on the dissection plates, replica-plate the crosses of the pJS801 transformants to strain 5-4 onto SC-ura, SC-leu, and YPD plates at 30°C to determine segregation of *URA3* and *LEU2*. For the crosses of the pJS801, pRS316, and pRS426 transformants to strain 5-3, replica-plate to SC-ura to monitor the plasmid segregation/stability and to SC-trp to determine centromere segregation.
2. Score the fraction of cells that remained Ura^+ after nonselective growth in YPD.

MATERIALS

Note: Amounts provided are the requirements for each pair.

Day 1 1 YPD plate

Day 3
1 culture tube containing 5 ml of YPD
Erlenmeyer flask containing 100 ml of YPD

Day 4
Materials for Techniques and Protocols #1, High-efficiency Transformation of Yeast
6 SC-ura plates

Day 6
3 SC-ura plates
3 SC-leu plates

Day 7
3 YPD plates
3 SC-ura plates

Day 8
1 SD plate
1 SD+his+leu plate

Day 9
2 YPD plates
6 culture tubes, each containing 5 ml of YPD

Day 10 18 culture tubes, each containing 5 ml of sterile H_2O
12 YPD plates
12 culture tubes, each containing 2 ml of sporulation medium
2 sporulation plates

Day 13 10 YPD plates for tetrad dissection
6 SC plates
6 SC-ura plates

Day 15 6 SC-ura plates
3 SC-leu plates
6 YPD plates
3 SC-trp plates

REFERENCES

Becker D.M. and Guarente L. 1991. High-efficiency transformation of yeast by electroporation. *Methods Enzymol.* **194:** 182–187.

Costanza M.C. and Fox T.D. 1988. Transformation of yeast by agitation with glass beads. *Genetics* **120:** 667–670.

Fox T.D., Sanford J.C., and McMullin T.W. 1988. Plasmids can stably transform yeast mitochondria lacking endogenous mtDNA. *Proc. Natl. Acad. Sci.* **85:** 7288–7292.

Hadfield C., Jordan B.E., Mount R.C., Pretorius G.H.J., and Burak E. 1990. G418-resistance as a dominant marker and reporter for gene expression in *Saccharomyces cerevisiae*. *Curr. Genet.* **18:** 303–313.

Hashimoto H., Morikawa H., Yamada Y., and Kimura A. 1985. A novel method for transformation of intact yeast cells by electroinjection of plasmid DNA. *Appl. Microbiol. Biotechnol.* **21:** 336–339.

Heinemann J.A. and Sprague G.F. 1989. Bacterial conjugative plasmids mobilize DNA transfer between bacteria and yeast. *Nature* **340:** 205–209.

Hinnen A., Hicks J.B., and Fink G.R. 1978. Transformation of yeast. *Proc. Natl. Acad. Sci.* **75:** 1929–1933.

Ito H., Fukuda Y., Murata K., and Kimura A. 1983. Transformation of intact yeast cells treated with alkali cations. *J. Bacteriol.* **153:** 163–168.

Johnston S.A., Anziano P., Shark K., Sanford J.C., and Butow R.A. 1988. Transformation of yeast mitochondria by bombardment of cells with microprojectiles. *Science* **240:** 1538–1541.

Murray A.W. and Szostak J.W. 1983. Pedigree analysis of plasmid segregation in yeast. *Cell* **34:** 961–970.

Orr-Weaver T., Szostak J., and Rothstein R. 1981. Yeast transformation: A model system for the study of recombination. *Proc. Natl. Acad. Sci.* **78:** 6354–6358.

Sikorski R.S. and Hieter P. 1989. A system of shuttle vectors and yeast host strains designed for efficient manipulation of DNA in *Saccharomyces cerevisiae*. *Genetics* **122:** 19–27.

EXPERIMENT VI

Synthetic Lethal Mutants

One of the advances resulting from the yeast genome project is a complete set of deletion mutants. A consortium of labs collaborated to construct deletion mutations of every gene in an isogenic S288C background that contains *ura3Δ0 his3Δ1 leu2Δ0* alleles (http://www.sequence.stanford.edu/group/yeast_deletion_project/deletions3.html).

The *ura3Δ0* and *leu2Δ0* are complete deletions and the *his3Δ1* allele is a 187-bp internal deletion of the *HIS3* gene. The deletions of all other open reading frames were constructed by one-step gene replacements using polymerase chain reaction (PCR)-derived fragments and the G418 resistance ($G418^r$) cassette from the pFA6a family of plasmids. Each deletion was constructed in a diploid so that the results were heterozygous strains. Correct integrations were confirmed by both PCR and tetrad analysis. Two general classes of genes are defined by the deletions. Nonessential genes are defined by deletion mutants that are viable. Therefore, the tetrads from these strains contain four viable spores, two of which are $G418^r$. Essential genes are defined by mutants that produce two viable spores, neither of which are $G418^r$.

The deletion strains are available and can be purchased from two sources. The American Type Culture Collection (ATCC) has U.S. distributors and the European *Saccharomyces cerevisiae* Archive for Functional Analysis (EUROFAN) consists of European distributors. Complete sets with deletions of nonessential genes are available in haploids and diploids, and the complete set of deletions (essential and nonessential genes) is also available as homozygous or heterozygous diploids.

Precise deletions of all genes add two important tools to the repertoire of yeast genetics. First, they represent the ultimate in saturation mutagenesis. Once the mutants are screened for a phenotype, you can be convinced that all of the possible mutants have been identified, because you have screened the entire deletion set. Second, if you identify a mutant with an interesting phenotype, you assign a function to the gene and the identity is largely certain. In most cases, it is not necessary to clone and confirm the identity. However, since contamination can occur in the knockout strain collection, you may want to confirm that the strain identified actually bears a knockout allele of the expected gene. Since each knockout allele was tagged with a unique "bar code" and flanked with universal primers, identity can simply be confirmed by PCR and a single sequencing reaction.

The most common use of the deletion mutants is to screen for a particular phenotype (for example, drug resistance or sensitivity). At the University of Toronto, Charlie

Boone has devised a simple way to use the deletion mutants to identify genetic interactions with double mutants (Tong et al. 2001, 2004; Hartwell 2004). We use Charlie's scheme to identify synthetic lethal interactions between a *mad2::NAT* allele and the deletion mutants. The principle of synthetic lethality is that a double mutant is inviable when either single mutant is viable. If you have a deletion of your favorite gene *yfgΔ1* and the strain is viable, and a deletion of an interacting gene (*iagΔ1*) is also viable but the *yfgΔ1 iagΔ1* double mutant is inviable, then you have a synthetic lethal interaction. Before the development of the Charlie Boone method, plasmids were used to identify synthetic lethal mutants after random mutagenesis with ethylmethanesulfate. However, we use crosses, meiosis, and a powerful haploid selection to identify double mutants and determine synthetic lethality.

Confirmation of our putative synthetic lethal interactions is performed by random spore analysis. The principle of random spore analysis is explained in Experiment IV. The usual procedure is to construct diploid strains that are heterozygous for two recessive drug-resistance markers such as *can1* and *cyh2*. The diploids are sporulated and the spores are plated onto medium containing both cycloheximide and canavanine. The selection for two recessive markers assures that the cells growing on the medium are haploid. The innovation behind the query strain is a gene replacement that removes the *CAN1* gene, thereby generating a *can1* mutant, and replaces it with a P_{STE2}-$his5^+$ fusion. The *STE2* gene encodes the receptor for α factor, the mating pheromone produced by *MAT*α cells. It is both a haploid-specific gene (expressed only in haploids) and is *MAT***a** specific (expressed only in *MAT***a** cells). The starting strain is *MAT*α and has the P_{STE2}-$his5^+$ fusion. We have deleted the *MAD2* gene to generate a *mad2::NAT* allele (strain 6-2).

The deletions were made by one-step gene replacements in strain 6-1 and are G418 resistant. The nonessential knockout collection has approximately 5000 strains, each with a different mutation (*orf::G418*).

In this experiment, the instructors previously crossed strain 6-1 to each of the deletion mutants. The diploids were previously selected on synthetic complete (SC) medium containing NAT and G418, and the diploids were sporulated. Select the haploids on SC-his medium containing canavanine (can). This represents haploid *MAT***a** cells that have the deletion of the nonessential gene and have gone through meiosis. The brilliance of the strain is that you recover haploids that cannot mate because they are all *MAT***a**. Test the haploids sequentially for growth on SC-his-arg +can+G418 and SC-his-arg +can+G418+NAT. If synthetic lethality is present, you cannot recover the double mutant and you will see no growth on G418+NAT. (*Media notes:* All SC media are made with the standard formulations except that 5 mg/ml ammonium sulfate is replaced with 1 mg/ml monosodium glutamate as the nitrogen source. This modification assures that the G418 is effective in synthetic medium.)

The cells are transferred with 384 floating replicator pinning tools that have been purchased from V&P Scientific (http://www.vp-scientific.com/htdocs/).

The colonies will be provided on square "Omnitrays" available from Nalgene. The colonies are in a 384 format and 19 plates cover the entire collection of mutants of nonessential genes. The transfers are accomplished using a "colony copier" that allows you to orient the transfers appropriately. This is demonstrated in the lectures. In this experiment, we have performed the crosses, selected diploids, and sporulated them. You are provided with the KAc sporulation plates. After screening for synthetic lethal mutants, you will recover the sporulated diploids and retest them by two secondary tests to reduce false positives in the screen. One is a semiquantitive random spore analysis and the second is a quantitative random spore analysis.

STRAINS

6-1	BY4741	*MAT***a** *ura3Δ0 his3Δ0 leu2Δ0 met15Δ0*
6-2	2466-1	*MATα ura3Δ0 his3Δ0 leu2Δ0 lys2Δ0 cyh2 can1::*P_{STE2}*-his5*$^{+}$ *mad2::NAT*

Day 1

You are provided with two omnitrays containing sporulated diploids. Go to the station containing the 384-well pinning tool and the washing stations. There, two plastic trays contain 100 ml of water, one contains bleach, and three trays contain increasing amounts of water. The final tray is a Pyrex dish containing alcohol. Dip the tool into the tray, resting on the lid of the tray for each wash. It is *very* important that the pins are suspended in the water or bleach and are *not* touching the bottom. The pinning tool is washed in each tray for 1 minute *only*. After the final water wash, dip the tool in alcohol and flame. Rotate the tool 180°, dip in the alcohol, and flame again. *Be very careful not to light yourself on fire.*

1. Water 1 min
2. Water 1 min
3. Bleach 1 min
4. Water 1 min
5. Water 1 min
6. Water 1 min

ETHANOL Flame

Rotate 180°

ETHANOL Flame

TOP

Remove the lid and place the copier over it with the plate oriented as shown above. Note that the notched edge faces down.
Place the copier over the plate so that the pin guide with four holes is at the bottom.

The copier has two sides. One side, the correct side, has four holes:

AB

CD

The correct side has 16 holes:

The replicator guide pins must go into both holes in the C position (see above), one on either side of the copier. Carefully pick up the cells, remove the copier from the master plate, and *replace the lid*. Remove the lid from the plate that will be the copy (SC-his-arg ser+can+ 2 mM 3AT+G418), cover with the second copier, and pin the cells using the guides in the guide hole labeled C.

Wash the pinning tool through the washes as described above. Incubate the plates at 30°C. Wrap the KAc plates in parafilm and refrigerate at 4°C.

Day 3

Pin the cells to SC-his-arg +can+G418 and SC-his-arg +can+G418+NAT. Wash and sterilize the pinning tool through the washes as described above. Incubate the plates at 30°C.

Day 4

Compare the growth of colonies on the SC-his-arg +can+G418 and SC-his-arg-+can3AT+G418+NAT plates. Score for colonies that grow on the SC-his-arg +can+G418 plate but *not* on the SC-his-arg +can+G418+NAT plate. They will have a position on the plate. The rows in the 384 format are A–P and the columns are 1–24 when observed in the standard position for the plates, as described above. Therefore, the position of the strain is "plate, row, column." For example, if you have plate 4 and a colony grows on the SC-his-arg +can+G418 plate but not on the SC-his-arg +can+G418+NAT plate in row K, column 4, it is a candidate for a synthetic lethal combination (4-K-4). Instructors will help you to identify the mutants.

Go to the Web site http://bud02.micr.virginia.edu:8080/yscgenomics. Click on the link "EUROSCARF Deletion Strains (384-well format)." In the upper panel, query plate 4 row K column 4. The answer will be *cin1*.

Day 6

Remove the KAc plates (stored from day 1) from the refrigerator. Add 200 µl of sterile water into each well of a 96-well microtiter dish. Identify the putative mutants from the screen and *carefully* resuspend one quarter of the sporulated cells you are testing in the wells of rows 1, 4, 7, and 10. (Do not use all the spores because you will need more later.) For each culture, remove 20 µl for two serial tenfold dilutions. For example, if you have cells in position A-1, perform tenfold dilutions into positions A-2 and A-3. Put parafilm on the KAc plates and return them to the refrigerator. Frog the spores to SC-his-arg -ser+can+cyh, SC-his-arg +can+cyh+G418, and SC-his-arg +can+cyh+G418+NAT plates.

Day 8

Score the plates. Interesting mutants grow on the SC-his-arg +can+cyh and SC-his-arg +can+cyh+G418 plates, but not on the SC-his-arg +can+cyh+G418+NAT plate.

Day 10

Remove the KAc plates (stored from day 1) from the refrigerator. Add spores to 200 µl of sterile water and plate 50 µl onto an SC-his-arg -ser+can+cyh plate.

Day 12

Replica-plate to SC-his-arg +can+cyh+G418, SC-his-arg +can+cyh+NAT, and SC-his-arg +can+cyh+G418+NAT plates.

Day 13

Score the plates and record the data.

REFERENCES

Hartwell L. 2004. Genetics. Robust interactions. *Science* **303:** 774–775.

Tong A.H., Evangelista M., Parsons A.B., Xu H., Bader G.D., Page N., Robinson M., Raghibizadeh S., Hogue C.W., Bussey H., Andrews B., Tyers M., and Boone C. 2001. Systematic genetic analysis with ordered arrays of yeast deletion mutants. *Science* **294:** 2364–2368.

Tong A.H., Lesage G., Bader G.D., Ding H., Xu H., Xin X., Young J., Berriz G.F., Brost R.L., Chang M., Chen Y., Cheng X., Chua G., Friesen H., Goldberg D.S., Haynes J., Humphries C., He G., Hussein S., Ke L., Krogan N., Li Z., Levinson J.N., Lu H., Menard P., Munyana C., Parsons A.B., Ryan O., Tonikian R., Roberts T., Sdicu A.M., Shapiro J., Sheikh B., Suter B., Wong S.L., Zhang L.V., Zhu H., Burd C.G., Munro S., Sander C., Rine J., Greenblatt J., Peter M., Bretscher A., Bell G., Roth F.P., Brown G.W., Andrews B., Bussey H., and Boone C. 2004. Global mapping of the yeast genetic interaction network. *Science* **303:** 808–813.

Experiment VII

Gene Replacement

One of the most powerful and important techniques available for studies in yeast is gene replacement. This technique allows the replacement of a gene at its normal chromosomal location with an allele of that gene created in vitro, such that the only genetic difference between the initial strain and the final strain is that particular allele. Using this method, phenotypes conferred by null mutations or any other types of mutations made in a cloned gene can be analyzed. In theory and in practice, a cloned yeast gene can be changed and then recombined into the genome, precisely replacing the wild-type allele.

Determining the null phenotype for a gene is an essential step in understanding the function of that gene. First, it reveals whether the gene is essential for growth. Second, if the gene is not essential for growth, it allows study of strains completely lacking that particular function. To examine the phenotype conferred by a null mutation, gene replacement is generally done in diploid strains, in case the null mutant is inviable as a haploid. The inviability will be observed after tetrad analysis of a sporulated culture. If the gene is essential, viability will segregate 2:2 in the tetrads. If the gene is not essential, all four spores will be viable and two will be null mutants.

Experiment VII(a)

ONE-STEP GENE DISRUPTION

Traditionally, a one-step gene replacement (Rothstein 1983) has been done by transformation with a restriction fragment that contains the mutant allele and has ends homologous to the locus where integration is desired. However, to create the mutant allele fragment, the gene of interest had to be cloned, and then using convenient restriction sites, the gene, or a portion of it, was replaced by a DNA fragment encoding a selectable marker in yeast. Amberg et al. (1995), Lafontaine and Tollervey (1996), and Wach (1996) described a simpler and more precise method based on polymerase chain reaction (PCR). Today it is routine to do a variety of manipulations using PCR-based gene replacements. We use a similar strategy devised by Longtine et al. (1998).

PCR-mediated gene disruption is based on the fact that homologous recombination in yeast is very efficient with linear DNA fragments, and that only ~40 bp of homology is frequently sufficient for homologous recombination. However, some loci can be difficult to target with short homologous sequences, in which case, longer regions of homology flanking the marker can be synthesized by double-fusion PCR (Amberg et al. 1995). The availability of the entire *Saccharomyces cerevisiae* genome sequence in combination with PCR-mediated gene disruption has made it possible to create a null mutation of any gene in any strain. We use a plasmid pFA6a-kanMX6 that contains an expression cassette with the kanMX6 module. This contains the known kanamycin resistance open reading frame of the *Escherichia coli* transposon Tn903 fused to transcriptional and translational control sequences of the *TEF1* gene of the filamentous fungus *Ashbya gossypii*. Expression of the cassette in *S. cerevisiae* confers resistance to the aminoglycoside G418. We use "fusion" PCR primers that contain 20 bp at their 3′ end, homologous to sequences 5′ and 3′ of the kanMX cassette and 40 bp of either the 5′ or 3′ end of *PEP4* (see Fig. 1). There is limited homology between the kanMX module and sequences in yeast, which eliminates unwanted background of gene conversion events. This can occur between the selectable markers and mutant alleles in the chromosome when using homotypic genes.

Often, gene disruptions are carried out to obtain a null allele of a gene of interest. In this experiment, we disrupt the gene for a protease to make a strain that will be useful for biochemical experiments. One of the difficulties in doing biochemistry on yeast cell extracts is the abundance of proteases contained within the vacuole that are released when cells are broken open. A truly protease-deficient strain requires disruption of each of the different genes that encode vacuolar proteases. Fortunately, there is a way to eliminate most of the proteases in a single step. Vacuolar proteases are synthesized as inactive proenzymes that become active in the vacuole on proteolytic removal of the propeptide. The *PEP4* gene encodes the major processing protease, proteinase A. Disruption of *PEP4* prevents all of the other proteases from becoming active. A complication of this procedure is that proteinase B can also cleave propeptides and activate vacuolar proteases. In the absence of proteinase A, active proteinase B can continue to process vacuolar proteases for several cell divisions, but the self-activation of proteinase B is not sustained indefinitely. Once the activity of proteinase B falls below

Figure 1. One-step gene replacement.

a threshold, the cell becomes irreversibly protease negative. After we disrupt *PEP4*, we streak out the disruption strain for single colonies several times to dilute proteinase B and to obtain a protease-negative strain.

A different PCR method will be used to screen transformed colonies for a disrupted *PEP4* gene. In this procedure, screening is done on colonies directly, without the need of purifying DNA from yeast. Transformants that are genotypically *pep4* are then screened, using a plate assay, for carboxypeptidase Y activity. This peptidase is a vacuolar enzyme that requires proteinases A and B for activation.

STRAIN

7-1 BY4741 *MAT***a** *his3Δ1 leu2Δ0 ura3Δ0 met15Δ0*

PLASMID

pFA6a-kanMX6 A plasmid carrying the kanMX6 module that can be used as a template for PCR-mediated gene disruption.

PROCEDURE

Day 1

Streak strain 7-1 on a YPD plate. Incubate at 30°C.

Day 3

Pick a robust colony of strain 7-1 and use it to inoculate a 5-ml YPD culture. Incubate overnight at 30°C.

Day 4

Determine the cell density of the 5-ml strain 7-1 culture using the hemocytometer (see Appendix G, Counting Yeast Cells with a Standard Hemocytometer Chamber). Dilute cells to 5×10^6 cells/ml in an Erlenmeyer flask containing 50 ml of YPD and grow culture (for about 4 hours) for two additional divisions at 30°C.

Harvest the cells and transform them, following Techniques and Protocols #1, High-efficiency Transformation of Yeast, with approximately 1 μg of the *pep4::kanMX* PCR-generated DNA (a stock will be provided that was made via Techniques and Protocols #14, PCR-mediated Gene Disruption), using pFA6a-kanMX6 as a vector template, and the two primers:

5´PEP4DEL TGGTCAGCGCCAACCAAGTTGCTGCAAAAGTCCACAAGGC**CGGATCC CCGGGTTAATTAA**

3´PEP4DEL AATCGTAAATAGAATAGTATTTACGCAAGAAGGCATCACC**GAATTGAGC TCGTTTAAAC**

Also, be sure to carry out a transformation with *no* DNA, as a negative control. Plate each transformation onto one YPD plate. Incubate for 1 day at 30°C.

Day 5

Replica-plate each plate to YPD plates containing 200 μg/ml of G418 (YPD+G418).

Day 7

Pick 11 transformants and streak for single colonies on YPD+G418 plates. Use only three plates and streak 3–4 transformants per plate. Incubate for 2 days at 30°C.

Day 9

1. Pick one colony from each of the 11 transformants that were streaked on YPD+G418 (from day 7) and determine whether the *PEP4* gene has been disrupted using Techniques and Protocols #15, Yeast Colony PCR. Also, be certain to include a sample of strain 7-1 cells as a control. Save the plates. Amplify the DNA by using the following four oligos:

   5´PEP4 GGGAACAGAGTAAAGAAGTTTGGG

   5´TEF GTTCTCACATCACATCCGAAC

   3´TEF GGGCTAAATGTACGGGCGAC

   3´PEP4 AGGATAGGGCGGAGAAGTAAGAAAAG

2. While the PCR amplification is being carried out, prepare a 1.0% agarose gel.
   Check the PCR products from each colony by gel electrophoresis. Add 3 μl of agarose gel loading buffer to each PCR sample and load all of each sample on the gel. Be certain to include a DNA size marker. The wild-type *PEP4* gene should produce an approximately 1.5-kb band, whereas the *pep4::KanMX6* allele will produce two bands of 341 bp (the 5´ junction) and an approximately 1.0-kb fragment (the 3´ junction).
3. Streak out three of the *pep4::KanMX6* transformants on YPD+G418 and strain 7-1 onto YPD plates. Streak two strains per plate. Incubate for 3 days at 30°C.

Day 11

Patch colonies from days 7 and 9 and from strain 7-1 onto a single YPD plate. Incubate the plates overnight.

Day 12

Assay the colonies on the YPD plates using the APE protease plate assay as in Techniques and Protocols #9, Plate Assay for Carboxypeptidase Y. Note any differences in color development among colonies.

MATERIALS

Day 1 1 YPD plate

Day 3 1 culture tube containing 5 ml of YPD

Day 4 Materials for Techniques and Protocols #1, High-efficiency Transformation of Yeast
Erlenmeyer flask containing 50 ml of YPD
2 SC-ura plates

Day 7 3 YPD+G418 plates

Day 9 Materials for Techniques and Protocols #15, Yeast Colony PCR
~120 pmoles of each DNA oligo:

5´PEP4	GGGAACAGAGTAAAGAAGTTTGGG
5´TEF	GTTCTCACATCACATCCGAAC
3´TEF	GGGCTAAATGTACGGGCGAC
3´PEP4	AGGATAGGGCGGAGAAGTAAGAAAAG

Materials and equipment for a 1% agarose gel
2 YPD+G418 plates

Day 11 1 YPD plate

Day 12 Materials for Techniques and Protocols #9, Plate Assay for Carboxypeptidase Y

Experiment VII(b)

TWO-STEP GENE REPLACEMENT

The second type of gene replacement commonly used is a two-step method (Scherer and Davis 1979) in which one first integrates a plasmid that contains the mutant allele. When the integration occurs at the locus of interest, it results in a duplication of the region—one duplicate will be wild type and the other mutant—with the plasmid sequences in between. Homologous crossovers between the copies from the duplication will result in excision of the plasmid and loss of one of the two copies of the duplicated region. Depending on the exact location of the crossover, the copy left behind contains either the mutant or wild-type form of the gene (Fig. 2). These can be distinguished by phenotype and/or PCR analysis. Excision of the plasmid is done most conveniently if the selectable marker on the plasmid is URA3; one can simply select for those that have lost the plasmid sequences using 5-fluoro-orotic acid (5-FOA) plates.

Two-step gene replacement is the method of choice under the following two conditions:

1. If the desired mutation is not associated with a selectable phenotype, direct selection by the one-step method is not possible. This is the case for gene replacement with a temperature-sensitive allele of a particular gene. Using the two-step method, one can screen recombinants that have looped out the plasmid for those containing the mutant allele.
2. Using the two-step method, one can generate both the wild-type and mutant strains from the same initial transformant strain. This provides strains that are isogenic except in the gene of interest. A two-step gene replacement protocol is used in Experiment IX to switch mating-type alleles at the *MAT2* locus.

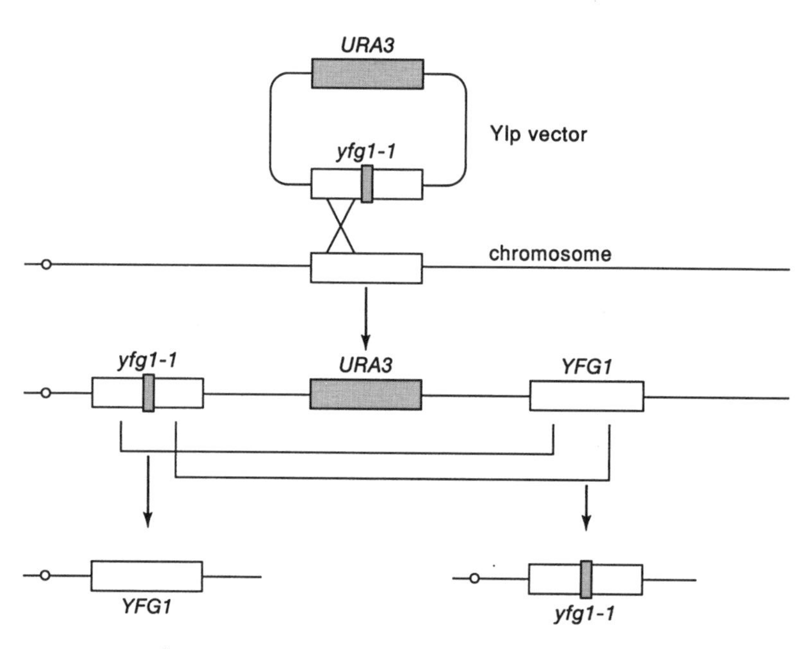

Figure 2. Two-step gene replacement. Integration of a plasmid by a single crossover results in a duplication of the *YFG1* locus. One copy contains the wild-type *YFG1* gene and the other copy contains the *yfg1-1* mutation. Strains that have excised the plasmid can be selected using 5-FOA medium. Depending on the location of the crossover, the excision will leave behind either *YFG1* or *yfg1-1*.

EXPERIMENT VII(C)

THE PLASMID SHUFFLE

In addition to analyzing phenotypes through gene replacement, it is possible to screen for mutant phenotypes when the gene of interest is on an autonomous plasmid, particularly when attempting to identify conditional lethal alleles of an essential gene, such as your favorite gene (*YFG*). Using the plasmid-shuffle technique (Sikorski and Boeke 1991), a plasmid containing *YFG* is mutagenized, the pool of mutagenized plasmids is transformed into yeast, and it is then screened for a conditional mutant phenotype in a strain that contains a *yfg* deletion mutation. This is done by transforming a strain that already contains (1) the wild-type gene of interest on an autonomous plasmid with a counterselectable gene (such as *URA3*) as the selectable marker (the counterselectable gene has both positive and negative selection and you can select cells that have the plasmid and cells that do not have the plasmid) and (2) a *yfg* null allele in the genome.

In such a strain, viability is dependent on the *YFG* encoded on the plasmid. A second plasmid with a different selectable marker is mutagenized (usually in vitro) and transformed into your strain. Replica-plating, to select against the plasmid containing *YFG*, identifies potential mutants. For *URA3* containing plasmids, the selection is done using 5-FOA plates under different conditions (e.g., temperature, media). Only cells that have lost the wild-type gene on the *URA3* plasmid can grow on the 5-FOA plates, and the phenotype conferred by the mutant plasmid will be expressed.

In this experiment, we look for temperature- and benomyl-sensitive alleles of the essential *NDC10* gene that encodes a component of the kinetochore (*NDC10=YFG*). Two libraries of mutagenized plasmids are provided, one mutagenized by hydroxylamine (Techniques and Protocols #7, Hydroxylamine Mutagenesis of Plasmid DNA) and the other by passage through *E. coli* strain XL-1 Red (Stratagene). The XL-1 Red strain is a *mutS, mutD, mutT* triple mutant. The three mutations almost completely eliminate DNA repair: *mutS* lacks error-prone mismatch repair, *mutD* lacks the 3´-5´ exonuclease of DNA polymerase III, and *mutT* lacks 8-oxodGTP hydrolysis. Plasmids are transformed into the XL-1 Red cells and then the strain is propagated overnight to accumulate mutations. The plasmids are recovered and amplified to give a library of mutagenized plasmids. The mutagenized plasmid pDB141 is a YCp *LEU2 NDC10* plasmid. Strain 7-2 has an *ndc10::TRP1* allele that is viable because it contains pRG68, a YCp *URA3 NDC10* plasmid that will be used for counterselection with 5-FOA as described above.

STRAIN

7-2 2404 *MATα ade2 trp1 leu2 ura3 his3 lys2-801 ndc10::TRP1* [pRG68]

PLASMIDS

pDB141 A YCp *LEU2* plasmid containing *NDC10*
pRG68 A YCp *URA3* plasmid containing *NDC10*

PROCEDURE

Day 1

Streak strain 7-2 onto a YPD plate. Incubate at 30°C.

Day 3

Pick a robust colony of strain 7-2 and use it to inoculate a 5-ml YPD culture. Incubate overnight at 30°C.

Day 4

Determine the cell density of the 5-ml strain 7-2 culture using the hemocytometer (see Appendix G, Counting Yeast Cells with a Standard Hemocytometer Chamber). Dilute cells to 5×10^6 cells/ml in 50 ml of YPD and grow the culture for two additional divisions (for about 4 hours) at 30°C. Harvest the cells and transform them, following Techniques and Protocols #1, High-efficiency Transformation of Yeast, with 100 ng of the mutagenized pDB141 plasmid (a stock will be provided that was made either via Techniques and Protocols #7, Hydroxylamine Mutagenesis of Plasmid DNA, or by passage in XL-1 Red cells). Be certain to include a no-DNA control. There will be a small modification to Techniques and Protocols #1, High-efficiency Transformation of Yeast. Modify step 16 by resuspending the transformed cells in 2.5 ml of sterile distilled H_2O (concentrated cells). Remove 0.4 ml of the concentrated cells and dilute this to 2.0 ml (diluted cells). Plate 0.2-ml aliquots onto HC-leu plates, using ten plates for the concentrated cells and another ten plates for the diluted cells. For the no-DNA control, plate a 0.2-ml aliquot of concentrated cells on one plate. Incubate all 21 plates at 23°C.

Day 9

Select ten transformation plates that ideally have 100–300 colonies each and determine the total number of colonies that will be assayed. Replica-plate each of the ten HC-leu plates onto 5-FOA plates. Keep the master plates (HC-leu). Incubate at 23°C.

Day 11

Determine the fraction of colonies that cannot papillate to FOA resistance. Replica-plate the FOA plates to two YPD plates and one YPD plate containing 15 μg/ml of benomyl (BEN). Incubate one YPD plate at 37°C, and the YPD and YPD+BEN plates at 23°C.

Day 12

Score the replica plates at 37°C for temperature-sensitive candidates. Pick as many as 12 temperature-sensitive colonies from the YPD plate incubated at 23°C and purify by streaking onto YPD at 23°C (use a maximum of three YPD plates).

Day 13

Score the BEN plates at 23°C for colonies that are benomyl sensitive and unable to grow or grow poorly. Pick as many as 12 candidates and purify by streaking on YPD at 23°C (use a maximum of three YPD plates). Note any benomyl-sensitive colonies that are also temperature sensitive.

Day 16

Retest the purified candidates for benomyl sensitivity on YPD+BEN and temperature sensitivity on two YPD plates, incubating one at 23°C and the other at 37°C .

Day 17

Score the plates at 37°C for temperature sensitivity. Score the benomyl plates at 23°C and keep these until tomorrow.

Day 18

Score the benomyl plates and determine the frequency of each class of mutants.

MATERIALS

Day 1 1 YPD plate

Day 3 1 culture tube containing 5 ml of YPD

Day 4 Materials for Techniques and Protocols #1, High-efficiency Transformation of Yeast

	Mutagenized pDB141 DNA 50 ml of YPD in a flask 21 HC-leu plates
Day 9	10 5-FOA plates Sterile velveteen pads
Day 11	20 YPD plates 10 YPD plates containing 15 μg/ml of benomyl Sterile velveteen pads
Day 12	3 YPD plates
Day 13	3 YPD plates
Day 16	3 YPD+BEN plates 6 YPD plates Sterile velveteen pads

EXPERIMENT VII (D)

CONSTRUCTING PROTEIN FUSIONS

Epitope tagging is a powerful technique that is used to analyze proteins in yeast. A protein fusion is made using a peptide sequence for which there are often commercially available antibodies. A number of different epitopes can be used: small peptide sequences that are highly antigenic such as the HA peptide from the hemagglutinin protein of influenza virus, a peptide (MYC) from the c-Myc protein, and the FLAG epitope. Fusions with small proteins also function as epitope tags and can have even greater utility. For example, GFP (green fluorescent protein) is used to localize fusion proteins in vivo and GST (glutathione *S*-transferase) is used to affinity-purify fusion proteins.

A number of different methods can be used to engineer protein fusions. The small peptides such as HA or MYC can be attached to either end of the protein or inserted in-frame in the coding sequence. Small protein fusions such as GFP or GST are usually made at either terminus of the protein. It is important to show, if at all possible, that the protein fusion to your favorite gene functions normally. This is best accomplished by complementing all mutant phenotypes associated with deletions of your favorite gene.

GFP FUSION

We will illustrate protein fusions using two different methods. The first will use a PCR-mediated one-step gene replacement (Longtine et al. 1998). We will generate a carboxy-terminal GFP fusion of an essential gene encoding a kinetochore protein. Kinetochores are localized near spindle poles through most of the cell cycle. The rationale is identical to the PCR strategy used in Experiment VII(A), except that the selectable marker is the *HIS3* homolog from the yeast *Schizosaccharomyces pombe*. The *S. pombe his5*$^+$ gene will complement *his3* mutants of *S. cerevisiae* and is used to reduce the possibility of gene conversion adding to the background of His$^+$ colonies after transformation. The primer design for the experiment is shown in Figure 3. Forty base pairs 5′ of the stop codon (see top arrow in Fig. 3) are used to make the 5′ end of the NDCTAG1 primer. In addition, the following sequence from the 5′ end of the *GFP* gene in plasmid pFA6a-GFP(S65T)-*HIS3*MX6 is added to the 3′ end of the NDCTAG1 primer:

CGGATCCCCGGGTTAATTAA

Forty base pairs 3′ of the stop codon are used to make the 5′ end of the NDCTAG2 primer. In addition, the following sequence from the 3′ end of the *S. pombe his5*$^+$ gene in plasmid pFA6a-GFP(S65T)-*HIS3*MX6 is added to the 3′ end of the NDCTAG2 primer:

GAATTCGAGCTCGTTTAAC

The sequences at the 3′ end of each primer are used to prime the PCR amplification from plasmid pFA6a-GFP(S65T)-*HIS3*MX6. The scheme is shown diagrammatically in Figure 4. The two 5′ ends of the primers direct the integration to make the precise GFP fusion at the carboxyl terminus of the Ndc10 protein.

To confirm the integration event, we will use a PCR primer from the 3′ end of the *NDC10* gene in combination with a primer from the *TEF1* promoter (primers 1 and 2 in Fig. 4). We will also use a primer from the *TEF1* promoter in combination with a primer from the 3′ flanking sequences of *NDC10* (primers 3 and 4 in Fig. 4). The PCRs

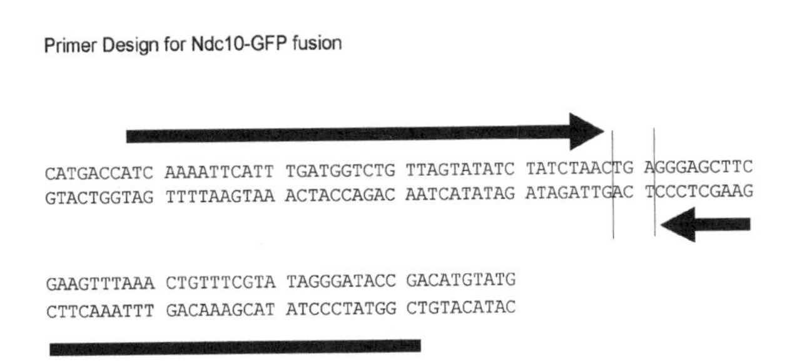

Figure 3. Primers for carboxy-terminal fusion to Ndc10p.

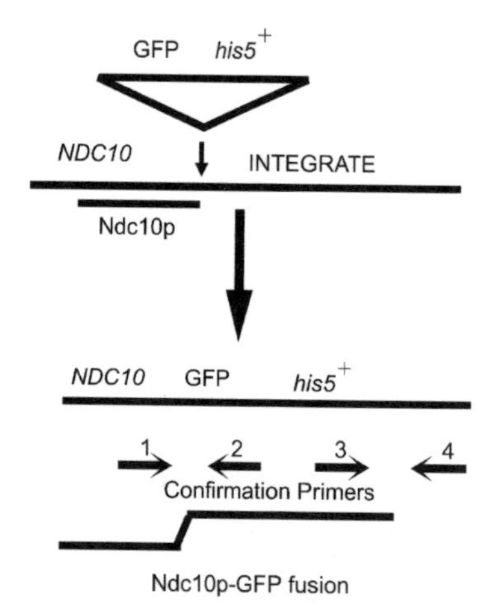

Figure 4. Ndc10p-GFP fusion.

are done together. A productive integration produces two PCR products of 1.2 and 1.4 kb. In the absence of integration, the *NDC10* gene remains intact and PCR produces a single product of 400 bp.

DEGRON FUSION

The second method will use integrative transformation with a YIp plasmid to generate an amino-terminal "degron" fusion to a different kinetochore protein Cep3p. Degron tagging is used to construct temperature-sensitive mutations in essential genes (Dohmen et. al. 1994). A degron is a complex amino-terminal fusion that targets the associated protein for proteolysis by the 26S proteasome. The fusion has the design shown in Figure 5.

The *CUP1* promoter regulates the fusion gene so that expression is dependent on the presence of copper in the medium. The degron has a single ubiquitin protein at the extreme amino terminus that provides the initiation codon for translation (M). The ubiquitin is rapidly cleaved upon expression and the cleavage occurs immediately after the ubiquitin peptide to uncover an arginine (R) residue at the amino terminus of the resulting fusion protein. The arginine is a destabilizing residue for the N-end rule and targets the protein for degradation when internal lysine residues are exposed. A mutant dihydrofolate reductase protein from mouse (mDHFR^{ts}) unfolds at 37°C to expose

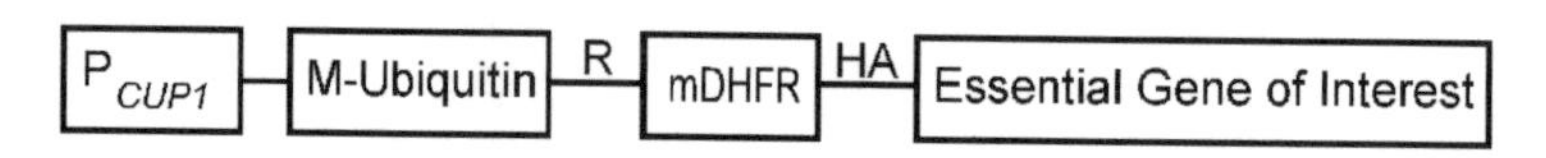

Figure 5. Degron-tagged genes.

lysines and target the entire fusion protein for proteolysis within the 26S proteasome. The 26S proteasome is highly efficient and, therefore, the mutant phenotype results from a complete lack of protein. This is easily confirmed because the fusion contains an internal HA epitope tag. The fate of the protein is determined by Western blots. The result is a fusion that generates a temperature-sensitive mutant that grows well at low temperatures (23–25°C) and is inviable at 37°C. Some degron-tagged mutants are somewhat "leaky" and grow slightly under restrictive conditions. This can usually be overcome by forcing the expression of excess Ubr1p, a ubiquitin protein ligase.

Degron-tagged temperature-sensitive alleles may be preferable to temperature-sensitive alleles constructed by in vitro mutagenesis. Degron tagging produces loss-of-function or null alleles. In vitro mutagenesis produces predominantly missense mutations and often there is residual protein activity, even at the restrictive temperature. The residual activity can sometimes confound the phenotype. There are two other utilities to the degron construct. The HA epitope can be used for localization studies and biochemical experiments with strains grown at the permissive temperature. In addition, the fusion protein is easily purified from cells grown at the permissive temperature using methotrexate-agarose beads because of the high affinity of DHFR for methotrexate.

We will construct a degron fusion to a different kinetochore protein, Cep3p. The integration with the YIp plasmid that produces the degron is shown in Figure 6. To confirm the integration, we will use a PCR primer to the region near the HA sequences at the 3′ end of the degron (primer A) in combination with a primer from within the 5′ end of the *CEP3* gene (primer B). The latter sequence is not contained within the *CEP3* fragment in the YIp plasmid but is unique in the chromosomal copy. This assures that the degron is fused to *CEP3*.

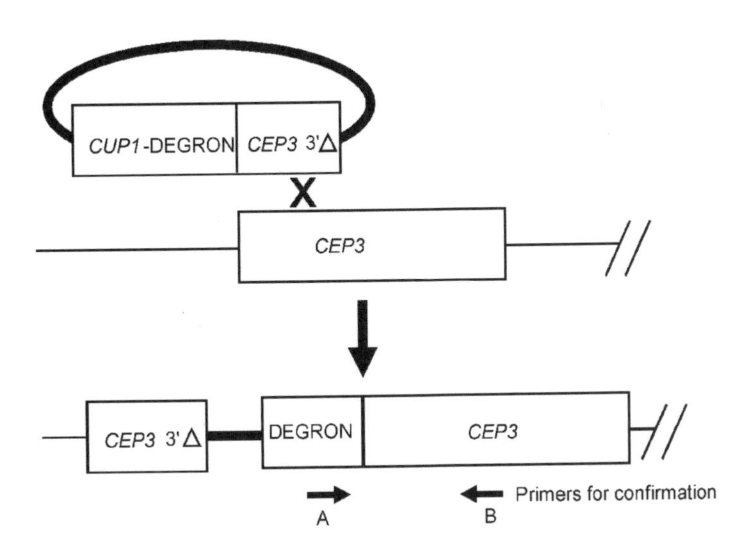

Figure 6. Degron-tagged *CEP3*.

STRAIN

7-3 BY4741 *MAT***a** *his3Δ1 leu2Δ0 ura3Δ0 met15Δ0*

PLASMID

pDB118 An integrating plasmid (YIp) containing *URA3* and the mouse DHFR fused to the 5′ end of *CEP3* and driven by the *CUP1* promoter.

PRIMERS

***NDC10*-GFP**

NDCTAG1
TCAAAATTCATTTGATGGTCTGTTAGTATATCTATCTAAC**AGTAAAGGAGAAGAACTTTT**

NDCTAG2
CGGTATCCCTATACGAAACAGTTTAAACTTCGAAGCTCCC**GAGCGCCCAATACGCAAACC**

***NDC10*-GFP Confirmation Primers**

Ndc10 confirm1
GGTAACAAGTCGTGGAGAGC

5′TEF
GTTCTCACATCACATCCGAAC

3′TEF
GGGCTAAATGTACGGGCGAC

Ndc10 confirm2
CAGGTCTACCAACTCAGTCTTC

Degron Tagging

Primer A
ACCTACCCATACGATGTTCCAG

Primer B
GACAGACCATATTAGTGCGTCCCAG

PROCEDURE

Day 1

Streak strain 7-3 for single colonies onto a YPD plate. Incubate for 2 days at 30°C.

Day 3

Pick an isolated colony and inoculate 5 ml of YPD liquid medium. Incubate overnight at 30°C.

Day 4

Determine the cell density using a hemocytometer (see Appendix G). Dilute cells to 5×10^6 cells per ml and proceed with a high-efficiency transformation following Techniques and Protocols #1, High-efficiency Transformation of Yeast. You will be provided with two DNA samples for transformation.

PCR products for GFP tagging: 100 µl containing approximately 8 µg of PCR product. The PCRs were previously performed using Techniques and Protocols #14, PCR-mediated Gene Disruption, using the plasmid pFA6a-GFP(S65T) as a template and primers NDCTAG1 and NDCTAG2. This DNA will be used for four separate transformations.

Bgl*II*-digested pDB118 for degron tagging: Approximately 1 µg of plasmid pDB118 previously digested with *Bgl*II. This DNA will be used for one transformation. Be sure to include a negative (no-DNA) control in the transformation.

Plate the transformation for GFP tagging onto SC-his plates. Plate the transformation for the degron tagging onto HC-ura plates containing 100 µM $CuSO_4$. Incubate the plates for 5 days at 23°C.

GFP Transformation

Day 7

Reclone the transformants (up to ten) on individual SC-his plates.

Day 10

Pick one colony from each of the transformants to identify GFP fusions. Use Techniques and Protocols #15, Yeast Colony PCR. Use four different primers, primers 1–4, in the reaction. Be sure to include cells from strain 7-3 as a control.

Separate the DNA by electrophoresis and identify the GFP fusions. Grow a 5-ml culture of cells in YPD liquid overnight.

Day 11

Inoculate 0.2 ml of the fresh overnight culture into 5 ml of YPD. Grow cells for 4 hours at 30°C to obtain cells that have reentered the cell cycle. Fix cells in formaldehyde according to Techniques and Protocols #12, Yeast Immunofluorescence. However, reduce the fixation to 15 minutes. Stain cells for both DNA with DAPI and visualize microtubules by antitubulin staining using CY3-conjugated secondary antibody.

Degron Transformation

Day 9

Replica-plate the transformants to HC-ura and to HC-ura containing 100 μM $CuSO_4$. Incubate the HC-ura plate at 37°C and the HC-ura containing 100 μM $CuSO_4$ at 23°C.

Day 10

Score the plates in the 37°C incubator and identify five temperature-sensitive colonies. Clone four candidate colonies to a single HC-ura plate containing 100 μM $CuSO_4$. Incubate at 23°C.

Day 14

Pick one colony from each of the transformants to identify degron fusions. Use Techniques and Protocols #15, Yeast Colony PCR. Use the two primers A and B in the reaction. Be sure to include cells from strain 7-3 as a control. Separate the DNA by electrophoresis and identify one strain containing the degron fusion. Grow a 5-ml culture of cells in HC-ura containing 100 μM $CuSO_4$ overnight.

Day 15

Inoculate 0.2 ml of the fresh overnight culture into 5 ml of HC-ura containing 100 μM $CuSO_4$. Grow cells for 6 hours at 23°C to obtain cells that have reentered the cell cycle. Remove 2.5 ml of cells, centrifuge, and wash twice with sterile water. Incubate the washed cells in HC-ura for 4 hours at 37°C. Return the rest of the culture to 23°C for 4 hours. Fix cells from both cultures in formaldehyde according to Techniques and Protocols #12, Yeast Immunofluorescence. However, reduce the fixation to 15 minutes. Wash the cells twice with 0.1 M potassium phosphate buffer (pH 7.5). Store the cells in a microfuge tube in 1 ml of 0.1 M potassium phosphate buffer (pH 7.5) in the refrigerator overnight.

Day 16

Stain cells for both DNA with DAPI according to Techniques and Protocols #12, Yeast Immunofluorescence, beginning at step 4. Visualize microtubules by antitubulin staining using CY3-conjugated secondary antibody.

MATERIALS

Day 1 1 YPD plate

Day 3 5 ml of YPD liquid medium

Day 4 Hemocytometer
PCR products for degron tagging (100 µl)
*Bgl*II-digested pDB118 DNA (25 µl)
5 SC-his plates
2 HC-ura plates containing 100 µM $CuSO_4$

Day 7 10 SC-his plates

Day 9 5 ml of liquid YPD medium
1 HC-ura plate
1 HC-ura plate containing 100 µM $CuSO_4$
Sterile velveteen pads

Day 10 1 HC-ura plate containing 100 µM $CuSO_4$
5 ml of liquid YPD medium
Materials for Techniques and Protocols #15, Yeast Colony PCR

Day 11 dH_20
5 ml of liquid YPD medium
Materials for Techniques and Protocols #12, Yeast Immunofluorescence

Day 14 5 ml of liquid HC-ura medium containing 100 µM $CuSO_4$
Materials for Techniques and Protocols #15, Yeast Colony PCR

Day 15 5 ml of liquid HC-ura medium containing 100 µM $CuSO_4$
5 ml of liquid HC-ura medium
37% formaldehyde
0.1 M phosphate buffer (pH 7.0)

Day 16 Materials for Techniques and Protocols #12, Yeast Immunofluorescence
CY3-conjugated secondary antibody

REFERENCES

Amberg D.C., Botstein D., and Beasley E.M. 1995. Precise gene disruption in *Saccharomyces cerevisiae* by double fusion polymerase chain reaction. *Yeast* **11:** 1275–1280.

Dohmen R.J., Wu P., and Varshavsky A.1994. Heat-inducible degron: A method for constructing temperature-sensitive mutants. *Science* **263:** 1273–1276.

Lafontaine D. and Tollervey D. 1996. One-step PCR mediated strategy for the construction of conditionally expressed and epitope tagged yeast proteins. *Nucleic Acids Res.* **24:** 3469–3471.

Longtine M.S., McKenzie A., Demarini D.J, Shah N.G., Wach A., Brachat A., Philippsen P., and Pringle J.R. 1998. Additional modules for versatile and economical PCR-based gene deletion and modification in *Saccharomyces cerevisiae. Yeast* **14:** 953–961.

Rothstein R.J. 1983. One-step gene disruption in yeast. *Methods Enzymol.* **101:** 202–211.

Scherer S. and Davis R.W. 1979. Replacement of chromosome segments with altered DNA sequences constructed in vitro. *Proc. Natl. Acad. Sci.* **76:** 4951–4955.

Sikorski R.S. and Boeke J.D. 1991. *In vitro* mutagenesis and plasmid shuffling: From cloned gene to mutant yeast. *Methods Enzymol.* **194:** 302–328.

Wach A. 1996. PCR-synthesis of marker cassettes with long flanking homology regions for gene disruptions in *S. cerevisiae. Yeast* **12:** 259–265.

EXPERIMENT VIII

Isolation of *ras2* Suppressors

Suppressor analysis is a powerful genetic approach for the study of genes in a common function or pathway. Suppressor analysis is a secondary approach that is used after an initial mutant with a well-characterized phenotype has been described. The suppressor hunt is designed to identify additional mutations that restore an aspect of the wild-type phenotype to the initial mutant. Suppressor mutations can lead to the identification of new genes with functions related to the initial mutated gene and can suggest interactions among proteins that participate in a common pathway. Alternatively, mutant phenotypes may arise when the balance between two genes, which operate in opposition (for example, a protein kinase and a protein phosphatase), is upset. The phenotype can be restored by compensatory mutations in the gene that encode the opposing protein. Such situations can occur when the first mutation is a null allele, in which case the second compensatory mutation "bypasses" the need for the function of the first gene. Such suppressors are referred to as "bypass suppressors."

High-copy suppression similarly uncovers genes in a common function or pathway. In this case, the suppression is achieved by overexpression of a wild-type copy of the suppressor gene. High-copy suppression can restore function by stabilizing productive protein complexes through mass action or by obviating the need for another gene (a null allele). The latter is called bypass suppression and has been particularly useful for ordering genes in signaling pathways (epistasis). High-copy suppression has gained in popularity as a result of ease of use, speed of use, and the fact that the process itself leads to cloning of the suppressor gene.

RAS1 and *RAS2* specify very similar G proteins that stimulate adenylate cyclase in yeast. cAMP is essential for growth and entry into the mitotic cell cycle. The components of the cAMP signal transduction pathway and their biochemical roles are well understood (see Broach 1991), in large part because of genetic analysis of cell growth regulation. In this experiment, we follow the approach of Cannon et al. (1986) to isolate suppressors of a *ras2* null allele in a classic suppressor hunt. In addition, we also use a 2μ-based genomic library to isolate high-copy suppressors of a *ras2* null allele.

Under most circumstances, *ras1* or *ras2* single mutants are viable and healthy; one functional RAS protein is sufficient to stimulate adenylylate cyclase. However, *ras2* mutants are defective specifically in growth on nonfermentable carbon sources (such as acetate or glycerol). The reason is that *RAS1* expression is repressed in such growth media. Therefore, a *ras2* null mutation causes a conditional phenotype. We exploit this phenotype to identify suppressors of *ras2*.

We have two goals in this experiment: First is the isolation of classical suppressors of the *ras2* defect. We will determine whether the suppressors are dominant or recessive as well as the number of complementation groups represented among the recessive suppressors. Second, we will isolate high-copy-number suppressors and the plasmids from these strains, confirm that the plasmid confers suppression, and by sequencing determine the genes carried by the high-copy-number plasmid.

STRAINS

8-1	FY86	*MATα ura3-52 leu2Δ1 his3Δ200*
8-2	DAY229	*MAT***a** *ura3-52 leu2Δ1 trp1Δ63 ras2Δ0::kan*^{r}
8-3	DAY230	*MATα ura3-52 leu2Δ1 his3Δ200 ras2Δ0::kan*^{r}

PROCEDURE

Day 1

Classic suppressor hunt. Instructors have streaked out your *ras2* mutant (strain 8-2 for groups 1–4; strain 8-3 for groups 5–8) on a pair of YPD plates.

Construct a patch master on the YPD with 30 to 40 isolated colonies, each approximately 1 cm^2. Incubate at 30°C. When you have finished making patches, give one of the plates of single colonies to a group using a mutant hunt strain of the opposite mating type in exchange for a plate of their strain. Refrigerate this plate for future use.

High-copy suppressor hunt. Start a 5-ml-YPD overnight culture of your *ras2* mutant. Incubate at 30°C.

Day 2

Classic suppressor hunt. Replica-plate the YPD master plate to one YPD plate and then to two YPAc plates. Replica-plate lightly to minimize background growth. Give the instructors one of the replicas for UV treatment (7500 μJ using a Stratalinker). Incubate both plates at 30°C.

High-copy suppressor hunt. First thing in the morning, dilute 500 μl of the now stationary liquid culture of your *ras2* mutant into 50 ml of YPD in a culture flask. Incubate for 6–7 hours until the cell density is approximately 2×10^7 per ml. Carry out nine transformations using 1 μg per transformation of the YEp *URA3* library DNA and one control transformation using 1 μg of YEp24 DNA. Plate on ten Cas-ura+acetate plates and incubate at 30°C.

Day 6

Classic suppressor hunt. Papillae (small colonies growing out of the patches) should be visible. Count the number of papillae observed per patch on both the mutagenized and nonmutagenized plates. Pick one papilla from each patch, choosing either the UV-treated plate or the non-UV-treated plate in each case, and patch on a fresh YPD master plate. Pick between 15 and 30 papillae. Pick carefully to avoid contamination with the surrounding cells. (Normally, one would purify the papillae by streaking for single colonies, but we omit that step to save time.) Number each patch and record which plate (spontaneous or UV-treated) the papilla came from. Include patches of the parent strain and the *RAS2* control strain 8-1 on the master plate, and incubate it at 30°C.

Inoculate 5 ml of YPD with the *ras2* strain you obtained from a neighboring group (strain 8-2 for groups 1–4; strain 8-3 for groups 5–8). Incubate overnight at 30°C.

Day 7

Classic suppressor hunt. Spread 200 µl of the cells from the 5-ml overnight culture evenly on an SC-his-trp plate; place the plate on a level surface and allow to dry.

Replica-plate the YPD master to two YPAc plates, to two YPD plates, and to the *ras2* lawn you have prepared on the SC-his-trp plate (be sure to do this step last). Incubate one YPAc/YPD pair at 30°C, the other at 37°C, and the mating plate at 30°C.

Day 8

Classic suppressor hunt. Score growth on the 30°C and 37°C plates. Save the 30°C YPD plate for use on day 12. Choose 12 revertants that you want to analyze in detail. Streak out the crosses of these 12 revertants to the *ras2* lawn on a SC-his-trp plate. In addition, streak out the cross of the *ras2* parent to the *ras2* lawn. You should be able to streak out four to six crosses per plate. Incubate these plates at 30°C.

High-copy suppressor hunt. Colonies should now be evident on your transformation plates. Pick eight colonies and inoculate 2-ml Cas-ura liquid overnight cultures, being careful not to pick any of the surrounding cells on the transformation plates. At the same time, use one quarter of a YPD plate to streak each for single colonies and incubate the plates at 30°C. Note that normally the colonies should be purified on a fresh plate before this step but that step is being omitted to save time.

Day 9

High-copy suppressor hunt. Use the 2-ml overnight cultures to isolate DNA from the strains carrying the putative high-copy suppressors using Techniques and Protocols #3C, A Ten-minute DNA Preparation from Yeast. Transform competent *Escherichia coli* with 5 µl and plate on LB+amp plates.

Day 10

Classic suppressor hunt. To test the suppressors for dominance or recessiveness, pick two isolated (diploid) colonies from each SC-his-trp streak and frog to YPAc and YPD plates. Include a colony of the cross of the *ras2* parent to the *ras2* lawn. Incubate at 30°C.

High-copy suppressor hunt. Start 5-ml LB+amp overnights from one colony of each putative high-copy suppressor from your *E. coli* transformations. Incubate at 37°C. Replica-plate the YPD streaks of your putative suppressors to two 5-FOA plates to select for cells that have lost the putative high-copy suppressor plasmid. These will be used to confirm that suppression is dependent on the presence of the plasmid.

Day 11

High-copy suppressor hunt. Miniprep plasmid DNA of your putative high-copy suppressors and transform the plasmid DNA into your *ras2* mutant using Techniques and Protocols #2, "Quick and Dirty" Plasmid Transformation of Yeast Colonies. Plate one half of each transformation on Cas-ura plates, the other half on Cas-ura+acetate plates and incubate at 30°C. Include a control transformation of 1-µg YEp24 DNA and plate one half of the transformation on a Cas-ura plate and the other half on a Cas-ura+acetate plate.

Day 12

Classic suppressor hunt. Score growth on the YPD and YPAc plates to determine which suppressors are dominant and recessive. Go back to the YPD plate you saved on day 8 and prepare a streak master on YPD, containing parallel streaks of the *ras2* parent and eight to ten recessive suppressor strains (see the streak template in Appendix D, Templates for Making Streak Plates). Incubate at 30°C.

High-copy suppressor hunt. To identify the suppressors that carry the *RAS2* gene, set up the following polymerase chain reactions (PCRs) using as templates the miniprep DNA of each of your putative high-copy suppressors:

1 µl of a 1:10 dilution of miniprep DNA
5 µl of 10x *Taq* buffer
5 µl of 2 mM dNTP mix
5 µl of 5 µM primer DAo-RAS2-1
5 µl of 5 µM primer DAo-RAS2-2
1 µl of *Taq*
28 µl of H_2O

Use the following PCR parameters: 94°C for 4 minutes, and then 25 cycles of 94°C for 1 minute, 55°C for 1 minute, and 72°C for 1.5 minutes, followed by 72°C for 20 minutes, and a 4°C soak.

Load 5 µl of each PCR on a 1% agarose gel with a *RAS2*-positive control PCR provided by the instructor. If the suppressor carries the *RAS2* gene, you can expect to see an approximately 1.5-kbp PCR product. Those plasmids that do not carry the *RAS2* gene will be sent for sequencing using the primers DAo-YEp24-1 and DAo-YEp24-2, which will identify the ends of the library inserts in the plasmid.

Replica from the 5-FOA plates from day 10 to two YPAc plates and incubate at 30°C.

Day 13

Classic suppressor hunt. To set up complementation group analysis of your recessive suppressors, replica-plate the streak master from day 12 to two YPD plates, label the streaks, and exchange one streak plate for one from a group that started their suppressor hunt with a strain of the opposite mating type. Incubate the plates at 30°C.

Day 14

Classic suppressor hunt. To do the matings for the complementation group analysis, replica-plate the pair of streak plates perpendicular to one another on a fresh YPD plate. Incubate at 30°C.

Day 15

Classic suppressor hunt. To select the diploids for the complementation test, replica-plate the crossed streaks from day 14 to an SC-his-trp plate. Incubate at 30°C.

Day 16

Classic suppressor hunt. Pick cells that grew on the SC-his-trp plate at the streak intersections and use them to prepare a patch master on an SC-his-trp plate.

Day 17

Classic suppressor hunt. Replica-plate the day-16 patch plate (SC-his-trp) to YPAc and YPD. Incubate at 30°C.

High-copy suppressor hunt. Score the replica plates onto YPAc from day 12 of the strains that have lost the putative high-copy suppressor plasmid. If these strains grow on YPAc, suppression is not plasmid dependent and the plasmids that were in these strains are not of interest.

Score the ability of the putative high-copy suppressors to allow growth of the *ras2* mutant on acetate medium. Note that the YEp24 control should have yielded colonies on Cas-ura but not on Cas-ura+acetate. If the candidate plasmid is, in fact, a high-copy suppressor, it should allow growth on acetate medium in all of the transformed cells. By now, you may have the sequence for the ends of the inserts in your suppressor plasmids.

Use the *Saccharomyces* Genome Database (http://www.pathway.yeastgenome.org/) to identify the genomic locus and the potential genes in that region that could be responsible for the suppression. In your own lab, individual open reading frames would then be subcloned into a 2μ plasmid and retested for high-copy suppression of the *ras2* allele to identify the actual suppressing gene.

Day 18

Classic suppressor hunt. Score the growth of the diploids on the plates from day 17. What phenotype do you expect from suppressors that fail to complement? Are any of your mutants in the same complementation group as those from your neighbors? How many complementation groups do you have?

MATERIALS

Day 1

2 YPD plates with single colonies of a *ras2* strain
1 YPD plate
Toothpicks
5 ml of YPD

Day 2

1 YPD plate
2 YPAc plates
1 sterile velveteen pad
50 ml of YPD
9 μg of Carlson 2μ library DNA
1 μg of YEp24 DNA
1.4 ml of 100 mM lithium acetate
2.4 ml of 50% PEG3350
360 μl of 2 mg/ml single-stranded DNA (ssDNA)
2.5 ml of sterile H_2O
10 Cas-ura+acetate plates

Day 6

1 YPD plate
Toothpicks
5 ml of YPD

Day 7

2 YPD plates
2 YPAc plates
1 SC-his-trp plate
1 sterile velveteen pad

Day 8

3 SC-his-trp plates
Toothpicks
16 ml of Cas-ura
8 sterile culture tubes
2 YPD plates

Day 9

8 1.5-ml microfuge tubes
1.6 ml of 2% Triton X-100/1% SDS/100 mM NaCl/TE buffer (10 mM Tris [pH 8.0]/1 mM EDTA)
1.6 ml of 1:1 phenol:chloroform
2.4 g of acid-washed glass beads
1.6 ml of competent *E. coli* cells
8 LB+ampicillin plates

Day 10

Toothpicks
1 YPAc plate
1 YPD plate
1 sterile 96-well microtiter dish
5 ml of sterile H_2O
40 ml of TB+ampicillin liquid medium
2 FOA plates

Day 11

24 microfuge tubes
800 μl of 50 mM glucose/10 mM EDTA/25 mM Tris (pH 8.0)
1.6 ml of 0.2 N NaOH/1% SDS
1.2 ml of 3 M potassium/5 M acetate
4 ml of phenol
6 ml of 100% ethanol
400 μl of TE buffer (pH 8.0) + 10 μg/ml of RNase A
200 μl of 2 M lithium acetate
800 μl of 50% PEG3350
7.7 μl of β-mercaptoethanol
24 μl of 10 mg/ml ssDNA
800 μl of sterile H_2O
9 Cas-ura+acetate plates
9 Cas-ura plates
1 μg of YEp24 plasmid DNA

Day 12

1 YPD plate
Toothpicks

40 μl of 10x *Taq* buffer
40 μl of 2 mM dNTP mix
40 μl of 5 μM primer DAo-RAS2-1 (5´-ccc caa cat ctt aag ttt tag ccg-3´)
40 μl of 5 μM primer DAo-RAS2-2 (5´-gct tgt tat tcc agg tgg aac acc-3´)
8 μl of *Taq*
8 PCR tubes
1% agarose gel
5 μl of *RAS2* control PCR
10 μl of 1-kb ladder DNA size markers
YEp24 sequencing primer 1: DAo-YEp24-1 (5´-ctt gga gcc act atc gac tac gcg-3´)
YEp24 sequencing primer 2: DAo-YEp24-2 (5´-cac ctg tgg cgc cgg tga tgc cgg-3´)
2 YPAc plates

Day 13
2 YPD plates
1 sterile velveteen pad

Day 14
1 YPD plate
1 sterile velveteen pad

Day 15
1 SC-his-trp plate
1 sterile velveteen pad

Day 16
1 SC-his-trp plate
Toothpicks

Day 17
1 YPAc plate
1 YPD plate
1 sterile velveteen pad

Thanks to Rey Sia for designing the classic suppressor hunt experiment and Bob Deschenes for help in designing the high-copy suppressor hunt experiment.

REFERENCES

Broach J.R. 1991. *RAS* genes in *Saccharomyces cerevisiae:* Signal transduction in search of a pathway. *Trends Genet.* **7:** 28–33.

Cannon J.F., Gibbs J.B., and Tatchell K. 1986. Suppressors of the *ras2* mutation of *Saccharomyces cerevisiae. Genetics* **113:** 247–264.

Experiment IX

Manipulating Cell Types

Saccharomyces cerevisiae is capable of undergoing sexual reproduction. Haploid cells are able to mate with one another, generating diploid cells. The diploid cells are then capable of undergoing meiosis to regenerate haploid cells. The ability to mate is determined by a cell's mating type. Cells of the same mating type are unable to mate, whereas cells of opposite mating type are able to mate. Mating type is determined by expression of alleles of the *MAT* locus. Cells of one mating type have the *MAT***a** allele and are designated **a** cells; cells of the opposite mating type have the *MAT*α allele and are designated α cells. Therefore, **a** cells are capable of mating with α cells, giving rise to diploids that contain both *MAT***a** and *MAT*α alleles. The resulting diploids are designated **a**/α.

The stability of mating type classifies strains into two groups, heterothallic and homothallic. Heterothallic strains have a stable mating type, whereas homothallic haploid strains are capable of changing mating type. Therefore, a colony derived from a homothallic cell will contain both **a** and α cells as well as diploid cells that arose from matings between **a** and α cells.

Spore-to-spore crosses between heterothallic and homothallic strains demonstrated that the distinction between homothallism and heterothallism segregates as a single genetic locus designated *HO* (Winge and Roberts 1949; Takahashi et al. 1958). It was shown that a homothallic cell switches from one mating type to the other early in its cell lineage (Strathern and Herskowitz 1979).

A variety of genetic analyses demonstrated the presence of two other loci involved in homothallism (Takahashi 1958; Takano and Oshima 1967). These loci are now designated *HML* and *HMR* and are found with the *MAT* locus on chromosome III (Fig. 1). It has been shown that *HML* and *HMR* contain silent copies of the information expressed from the *MAT* locus (Hicks and Herskowitz 1977; Klar et al. 1979; Strathern et al. 1979). In general, *HML* has the α-cell-type information and *HMR* has the **a**-cell-type information. *HO* has been shown to encode a site-specific endonuclease that cuts

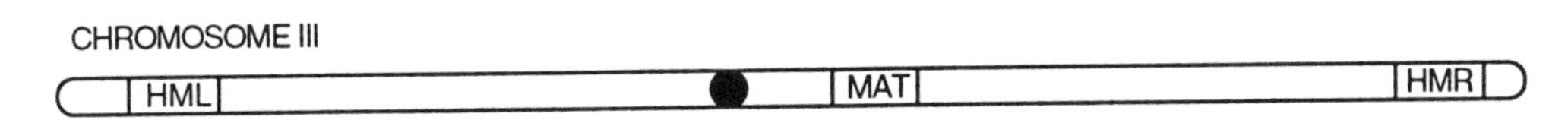

Figure 1. Diagrammatic representation of the silent loci *HML* and *HMR* in relation to the *MAT* locus on chromosome III.

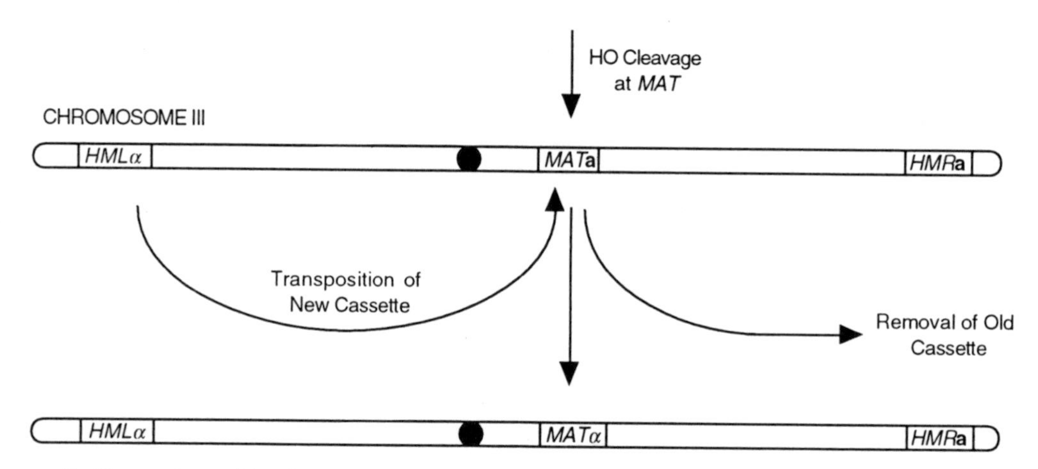

Figure 2. General model of mating type switching by the cassette model. In this case, α information replaces **a** at the *MAT* locus. (Adapted from Herskowitz and Oshima 1981, p. 194.)

only within the *MAT* locus. The double-strand break that occurs allows for recombination between the *MAT* locus and either *HML* or *HMR*. The recombination event, in general, occurs between the *MAT* locus and the silent locus containing information for the opposite mating type from *MAT*. Since this recombination event is not reciprocal, the information from the silent locus remains intact whereas the information at the *MAT* locus changes. These observations led to the cassette model for mating type switching (Fig. 2).

The ability to generate isogenic cell lines of different mating types is an important tool that facilitates the characterization of novel mutations. In this experiment, we use two different approaches to change the mating type of cells. We use *HO* to promote a switch by the natural homothallic pathway, and we use a two-step gene replacement technique to substitute a *MAT* allele from a plasmid for the starting *MAT* allele.

EXPERIMENT IX(A)

USING HO TO GENERATE DIPLOIDS

In this section, a haploid strain is transformed with a plasmid containing *HO*. Expression of the *HO* gene allows the cells to change mating type. Cells of opposite mating type can mate to generate diploids. In **a**/α cells, the *HO* gene is repressed so that the **a**/α cell type is stable. Cells that have lost the plasmid can be identified and tested for having changed mating type. This procedure allows rapid purification of isogenic **a** and α haploid, and **a**/α diploid cell types.

STRAINS

9-1	NE2	*MATα ura3-52, leu2-3,112*
9-2	AAY1017	*MATα his1*
9-3	AAY1018	*MAT***a** *his1*

Note: Most laboratory strains, including these three strains, are heterothallic because of an *HO* mutation, but that mutation is rarely listed in the strain genotypes. Strains 9-2 and 9-3 are useful for testing mating types, but the genetic background is unknown. Therefore, these strains should never be used for genetic crosses other than mating-type testing.

PLASMID

pCY204 *HO* in YCp50. Selectable marker is *URA3* (Fig. 3).

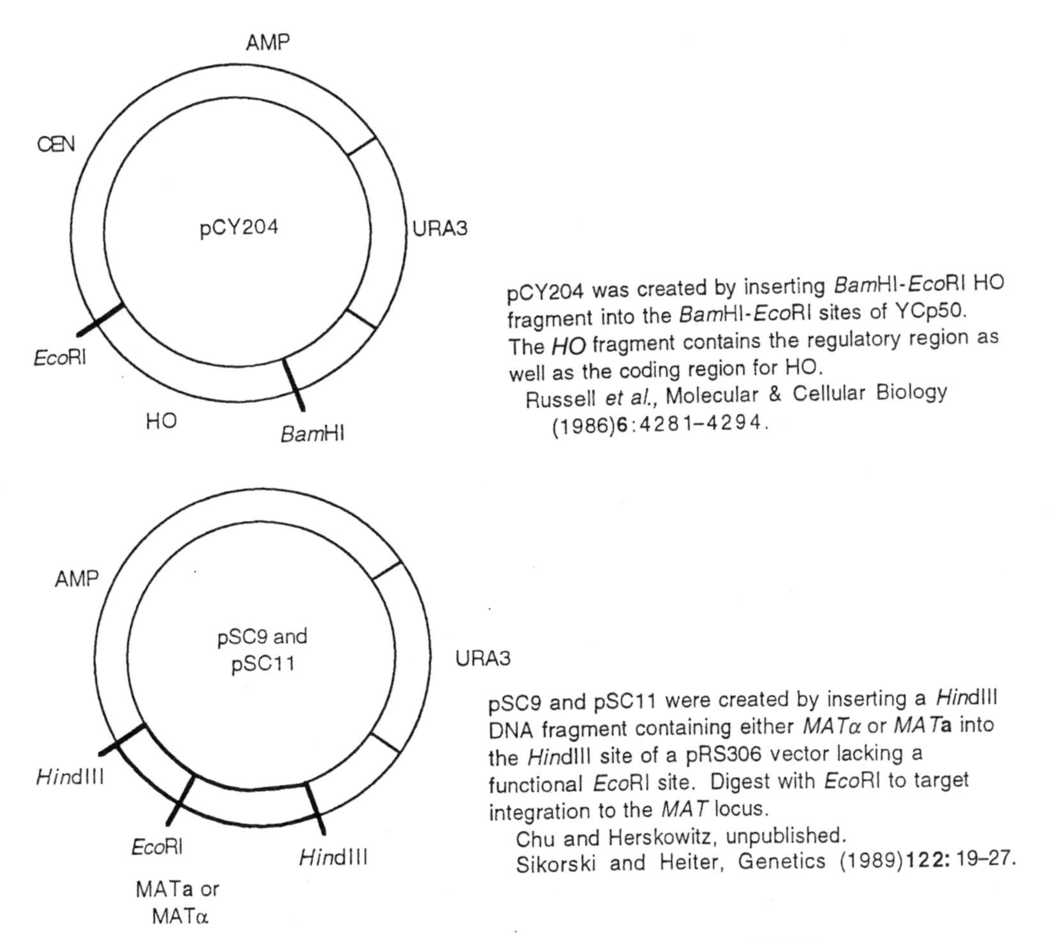

Figure 3. Plasmids for Experiments IX(A) and IX(B).

PROCEDURE

Day 1

Streak strain 9-1 for single colonies on a YPD plate and incubate at 30°C.

Day 3

In the morning, inoculate 5 ml of YPD with a colony of strain 9-1 and incubate at 30°C. Store the YPD plate from above at 4°C for use on day 13.

Day 4

Transform strain 9-1 following Techniques and Protocols #1, High-efficiency Transformation of Yeast, with uncut plasmid pCY204 (from your instructor). Remember to perform a no-DNA control transformation as well. Plate one 200-µl aliquot of transformed cells on an SC-ura plate and incubate at 30°C. Do the same for the no-DNA control.

Day 7

Streak four transformants for single colonies on SC-ura and grow at 30°C.

Day 9

Inoculate 5 ml of YPD with a single colony from each individual transformant and grow at 30°C. This allows growth of cells that have lost the plasmid.

Day 10

Determine cell density using a hemocytometer (see Appendix G, Counting Yeast Cells with a Standard Hemocytometer Chamber). Dilute cells appropriately in sterile distilled H_2O and plate 10^6 and 10^5 cells in 100-µl volumes on 5-FOA plates. This will select against the *URA3* gene so that only cells that have lost pCY204 can grow. Store the remaining YPD cultures at 4°C for use on day 12.

Day 12

Make a patch master plate by patching nine 5-FOA^R colonies of each transformant onto a YPD plate (36 total). In addition, patch the parent strain 9-1. Spot 3 µl of each of the four intermediate cultures from day 10 onto the patch master plate. Incubate the YPD plate at 30°C. Inoculate two 5-ml YPD cultures with strains 9-2 and 9-3. Grow at 30°C with agitation.

Day 13

On an SD plate, spread 150 µl of strain 9-2 and 150 µl of sterile H_2O and allow to dry. Repeat this process with another SD plate using strain 9-3. Replica-plate the patch master plate onto the SD plates with each lawn. Use a fresh velvet for each plate to prevent contamination between plates. In addition, replica-plate the patch master plate onto a sporulation plate and an SC-ura plate.

Day 15

Score mating ability and growth on the SC-ura plate.

Day 17

Score sporulation ability by using a toothpick to transfer cells to a drop of water on a slide and looking at cells in a phase contrast microscope.

MATERIALS

Note: Amounts provided are the requirements for each pair.

Day 1 1 YPD plate

Day 3 1 culture tube containing 5 ml of YPD

Day 4 Materials for Techniques and Protocols #1, High-efficiency Transformation of Yeast
Uncut plasmid pCY204
Erlenmeyer flask containing 50 ml of YPD
2 SC-ura plates

Day 7 1 SC-ura plate

Day 9 4 culture tubes containing 5 ml of YPD

Day 10 Sterile dH_2O
8 5-FOA plates

Day 12 1 YPD plate
2 culture tubes containing 5 ml of YPD

Day 13 2 sterile velveteen pads
2 SD plates
1 sporulation plate
1 SC-ura plate

EXPERIMENT IX(B)

MATING-TYPE SWITCHING BY TWO-STEP GENE REPLACEMENT

Two-step gene replacement allows the replacement of one allele of a gene with another allele (see also Expt. VII). This technique is generally used to replace a wild-type allele with a mutant allele that has no selectable phenotype. The first step of this procedure involves integrating a plasmid sequence containing a selectable marker and the gene of interest. This type of integration leads to a duplication of the gene of interest, with the two genes separated by the plasmid sequence (Fig. 4A). The second step is loss of the plasmid sequence by homologous recombination among the duplicated regions. Loss of the plasmid is most easily accomplished if the marker used to select for integration can be selected against, for example, by selecting against *URA3* with 5-fluoro-orotic acid (5-FOA). Depending on where the recombination event occurs, the remaining copy will carry one allele or the other (Fig. 4B). In this experiment, we use two-step gene replacement to switch mating types (S. Chu and I. Herskowitz, unpubl.).

STRAINS

9-4	YSC006	*MATα ura3 ade2-1 trp1-1 can1-100 leu2-3,112 his3-11,15[psi⁺]GAL⁺*
9-5	YSC005	*MAT*a *ura3 ade2-1 trp1-1 can1-100 leu2-3,112 his3-11,15[psi⁺]GAL⁺*

PLASMIDS

pSC9	*MATα* in pRS306 selectable marker is *URA3* (Fig. 4A).
pSC11	*MAT*a in pRS306 selectable marker is *URA3* (Fig. 4A).

PROCEDURE

Day 1

Streak strain 9-4 or 9-5 for single colonies onto YPD plates and incubate at 30°C. Groups 1–4 use strain 9-4 and groups 5–8 use strain 9-5.

Day 3

In the morning, inoculate 5 ml of YPD with a single colony of strain 9-4 or 9-5 and incubate at 30°C. Store the YPD plate from above at 4°C for use on day 13.

Day 4

Transform strain 9-4 or 9-5 following Techniques and Protocols #1, High-efficiency Transformation of Yeast. Obtain digested plasmid DNA from the instructor for trans-

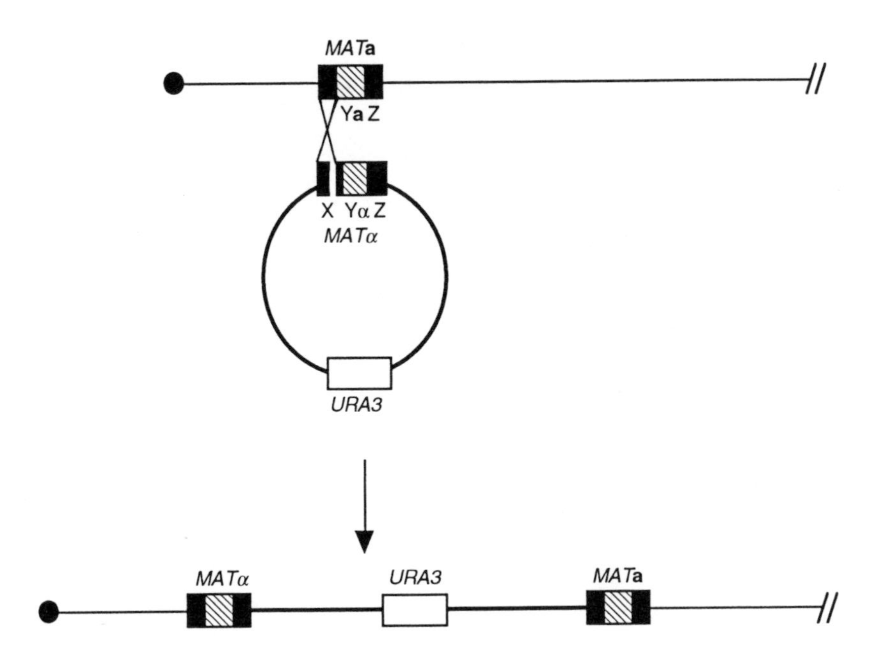

Figure 4A. Two-step gene replacement.

Step 1: Integration of plasmid by homologous recombination at duplicated regions, either X or Z. Unlike one-step gene replacements, the entire plasmid is integrated. *Note:* The relative order of the two alleles is determined by the position of the crossover. In this figure, sequences determining mating type occur 3′ of the crossover event.

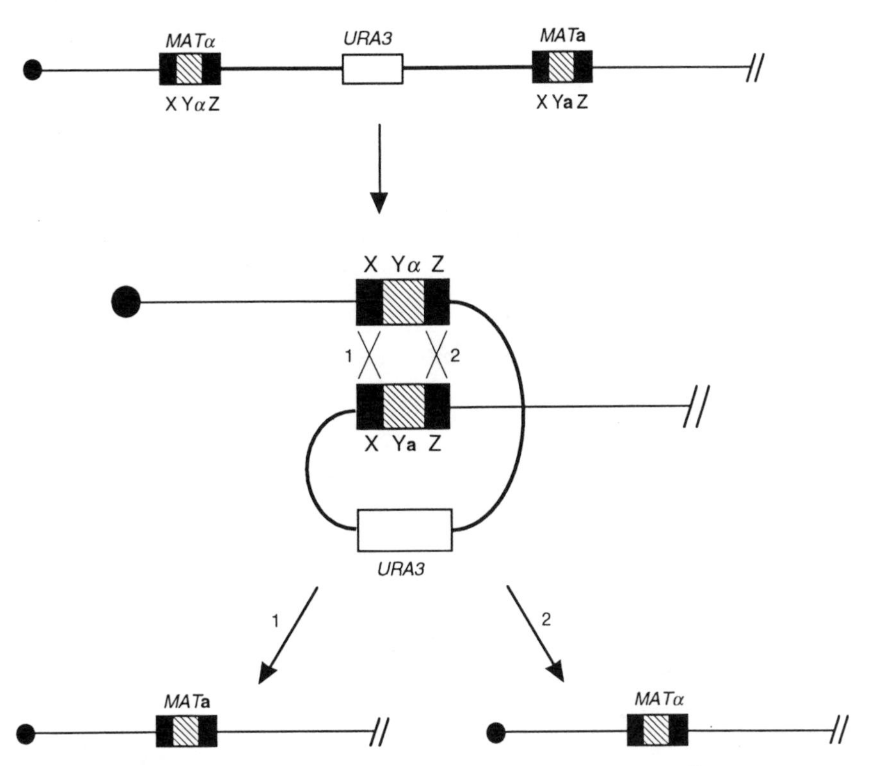

Figure 4B. Two-step gene replacement (continued).

Step 2: Selection against the integration (in this case, selecting against *URA3* by plating on 5-FOA) allows for "looping out" by homologous recombination between the duplicated regions of the *MAT* locus (X or Z). Depending on where recombination occurs (X or Z), the mating type will either be switched to *MAT***a** or reverted to the parent allele (*MATα*).

formation. Strain 9-4 will be transformed with pSC11 and strain 9-5 with pSC9. Remember to perform a no-DNA control transformation as well. Plate one 200-µl aliquot of transformed cells on an SC-ura plate and incubate at 30°C. Do the same for the no-DNA control.

Day 6

Streak four transformants for single colonies on SC-ura and grow at 30°C.

Day 9

Inoculate 5 ml of YPD with a single colony from each individual transformant and grow at 30°C.

Day 10

Determine cell density using a hemocytometer (see Appendix G, Counting Yeast Cells with a Standard Hemocytometer Chamber). Dilute cells appropriately in sterile distilled H_2O and plate 10^6 and 10^5 cells in 100-µl volumes on 5-FOA plates. This will select against the *URA3* gene that should be inserted between the duplicated regions of the *MAT* locus. 5-FOA^R arises through "looping out" of the *URA3* gene by homologous recombination. Store the remaining YPD cultures at 4°C for use on day 13.

Day 13

Make a patch master plate by patching nine 5-FOA^R colonies of each transformant onto a YPD plate (36 total). In addition, patch the parent strains 9-4 and 9-5. (Obtain one parent strain from the plate stored at 4°C on day 3, and the other parent from a neighboring group.) Spot 10 µl of each of the four intermediate cultures from day 9 onto the patch master plate. Incubate the YPD plate at 30°C.

Day 14

On an SD plate, spread 100 µl of strain 9-2 (from Expt. IX[A], day 13) and allow to dry. Repeat this process with another SD plate using strain 9-3. Replica-plate the patch master plate onto the SD plates with each lawn. Use a fresh velvet for each plate to prevent contamination between plates. In addition, replica-plate the patch master plate onto a sporulation plate and an SC-ura plate.

Day 16

Score mating ability and growth on the SC-ura plate.

Day 17

Score sporulation ability.

MATERIALS

Note: Amounts provided are the requirements for each pair.

Day 1 1 YPD plate

Day 3 1 culture tube containing 5 ml of YPD

Day 4 Materials for Techniques and Protocols #1, High-efficiency Transformation of Yeast
Digested plasmid pSC9 or pSC11 DNA
Erlenmeyer flask containing 50 ml of YPD
2 SC-ura plates

Day 7 1 SC-ura plate

Day 9 4 culture tubes containing 5 ml of YPD

Day 10 Sterile distilled H_2O
8 5-FOA plates

Day 13 1 YPD plate

Day 14 1 sporulation plate
1 SC-ura plate
2 SD plates
2 sterile velveteen pads

The method of switching mating type by two-step gene replacement was developed by Shelley Chu and Ira Herskowitz. Thanks to them for providing the methodology and materials for this procedure. Thanks also to Dana Davis for the initial write up of the experiment.

REFERENCES

Herskowitz I. and Oshima Y. 1981. Control of cell type in *Saccharomyces cerevisiae*: Mating type and mating-type interconversion. In *The molecular and cellular biology of the yeast* Saccharomyces: *Life cycle and inheritance* (ed. J.N. Strathern et al.), pp. 181–209. Cold Spring Harbor Laboratory, Cold Spring Harbor, New York.

Hicks J.B. and Herskowitz I. 1977. Interconversion of yeast mating types. II. Restoration of mating ability to sterile mutants in homothallic and heterothallic strains. *Genetics* **85:** 373–393.

Klar A.J.S., Fogel S., and Radin D.N. 1979. Switching of a mating-type a mutant allele in budding yeast *Saccharomyces cerevisiae. Genetics* **92:** 759–776.

Russell D.W., Jensen R., Zoller M.J., Burke J., Errede B., Smith M., and Herskowitz I. 1986. Structure of the *Saccharomyces cerevisiae HO* gene and analysis of its upstream regulatory region. *Mol. Cell. Biol.* **6:** 4281–4294.

Sikorski R.S. and Hieter P. 1989. A system of shuttle vectors and yeast host strains designed for efficient manipulation of DNA in *Saccharomyces cerevisiae. Genetics* **122:** 19–27.

Strathern J.N. and Herskowitz I. 1979. Asymmetry and directionality in production of new cell types during clonal growth: The switching pattern of homothallic yeast. *Cell* **17:** 371–381.

Strathern J.N., Blair L.C., and Herskowitz I. 1979. Healing of *mat* mutations and control of mating type interconversion by the mating type locus in *Saccharomyces cerevisiae. Proc. Natl. Acad. Sci.* **76:** 3425–3429.

Takahashi T. 1958. Complementary genes controlling homothallism in *Saccharomyces. Genetics* **43:** 705–714.

Takahashi T., Saito H., and Ikeda Y. 1958. Heterothallic behavior of a homothallic strain in *Saccharomyces cerevisiae. Genetics* **43:** 249–260.

Takano I. and Oshima. Y. 1967. An allele specific and a complementary determinant controlling homothallism in *Saccharomyces oviformis. Genetics* **57:** 875–885.

Winge Ø. and Roberts C. 1949. A gene for diploidization in yeast. *C.R. Trav. Lab. Carlsberg Ser. Physiol.* **24:** 341–346.

Isolating Mutants by Insertional Mutagenesis

There are two kinds of microtubules in yeast that have different functions in the life cycle. Nuclear microtubules, the major components of the intranuclear mitotic spindle, are required for chromosome separation. Cytoplasmic microtubules are required for nuclear positioning between the daughter cells and for nuclear fusion during mating (karyogamy). Mutants lacking cytoplasmic microtubules are viable whereas cells lacking nuclear microtubules are inviable, suggesting that the essential function of the microtubules is chromosome segregation.

Benomyl is a benzimidazole drug that is a potent inhibitor of microtubule assembly. At high concentration, the drug arrests cells at mitosis and at lower concentration, cells delay before anaphase in the cell cycle. The sublethal doses of benomyl are especially useful for genetic screens and two classes of mutants have been identified: One class of benomyl-sensitive mutants includes genes that affect the essential microtubule functions. The other class of mutants affects cell cycle regulation and the ability to arrest in the cell cycle in response to microtubule dysfunction.

INSERTIONAL MUTAGENESIS

In this experiment, we screen for benomyl-sensitive mutants using insertional mutagenesis. A library of yeast genes was constructed in a bacterial vector and the library was mutagenized with a transposon in *Escherichia coli* (Burns et al. 1994; Ross-Macdonald et al. 1999). A modified Tn3 transposon contains the *LEU2* gene from yeast and the *lacZ* gene from *E. coli* subcloned between the 38-bp inverted repeats that mark the end of the transposon. The transposon lacks all functions necessary for transposition (they are supplied in *trans*) and is referred to as a minitransposon or mTn3-*lacZ/LEU2*. The *lacZ* gene is adjacent to one of the 38-bp repeats and lacks the codon for initiating translation. Therefore, it is possible that some of the transpositions (approximately one-sixth) are in-frame *lacZ* fusions. The modified version of the *lacZ* gene is sometimes referred to as *lacZ′*. The transposon mutagenesis is accomplished by a two-step bacterial mating. The strain containing the library is successively mated to two strains that supply mTn3, the transposase and resolvase functions in *trans*. The result is that the library is recovered as a large number of plasmids (greater than 10^5), each of which contains a different transposition event. The plasmid DNA is recovered and digested with the 8-bp restriction enzyme *Not*I that excises the yeast inserts containing the mTn3-*lacZ/LEU2*. The DNA is transformed into a *leu2* yeast strain to produce one-step gene replacements. In this case, the "replacement" fragment is a randomly mutagenized region of the yeast genome.

Transposon mutagenesis has two important characteristics. The first is that the mutagen (the transposon) generates insertion mutations. Insertions into essential genes are generally lethal events and are not recovered if the transformation is done into haploid cells. This introduces a significant bias in the experiment because we primarily expect to recover mutations in nonessential genes. The second characteristic is that the insertion marks the mutated gene, providing a powerful and effective tool for molecular cloning and identification. We make use of the highly efficient recombination system in *Saccharomyces cerevisiae* to construct mutants. We exploit the sequenced genome to identify the genes.

There are two ways to identify the mutant. The first is to rescue the transposon and adjacent sequences as a plasmid. In this experiment, we use the second method and recover the flanking sequences by inverse polymerase chain reaction (PCR); the procedure is illustrated in Figure 1. The thin line represents mTn3-*lacZ/LEU2* transposon DNA and the thick line represents yeast DNA. Primers from the sequences at the *lacZ*

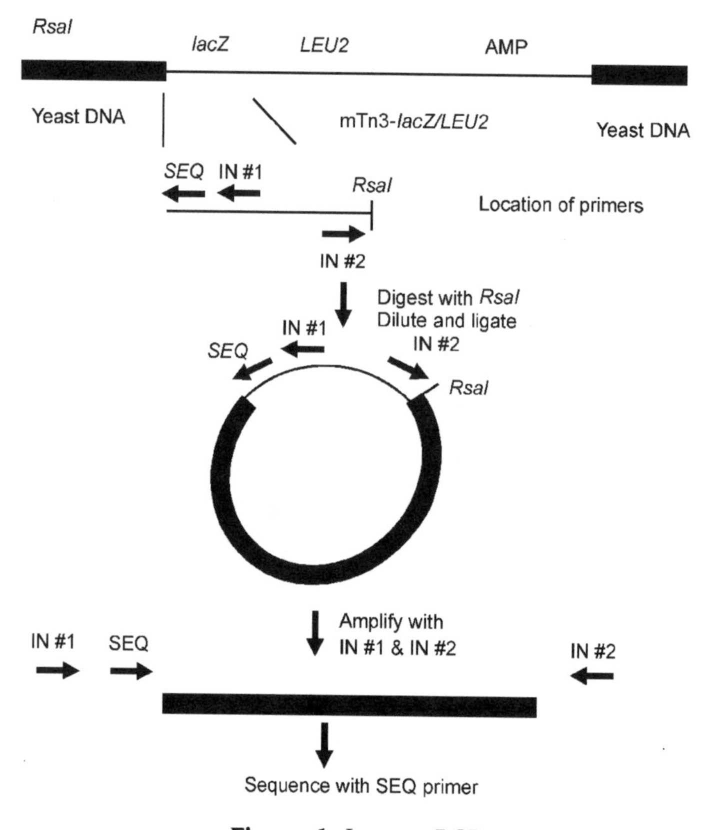

Figure 1. Inverse PCR.

end of the transposon are shown schematically as arrows. The restriction enzyme site for *Rsa*I, which recognizes a 4-bp sequence (GTAC), is shown. One site is adjacent to the IN #2 primer used for inverse PCR. The other *Rsa*I site is some unknown distance within the yeast DNA. The distance is likely to be small because of the high density of *Rsa*I sites in the genome. Yeast DNA is digested with *Rsa*I, diluted, and ligated. The dilution promotes intramolecular ligation and the formation of circles. The IN #1 and #2 primers are used to amplify a fragment that contains the yeast DNA sequences from the *lacZ´* end of the transposon to the flanking *Rsa*I site. The sequence of the DNA is determined using the SEQ primer to sequence the DNA fragment.

We use mTn3-*lacZ/LEU2* mutagenesis to screen for benomyl-sensitive mutants. A wild-type culture of cells is mutagenized and the surviving cells are plated onto HC-leu plates to select transformants. Colonies are replica-plated to HC-leu plates containing benomyl and screened for benomyl-sensitive mutants. Inverse PCR and DNA sequencing are then used to identify the integration sites in the benomyl-sensitive mutants.

STRAIN

10-1 BY4741 *MAT***a** *his3Δ1 leu2Δ0 ura3Δ0 met15Δ0*

PRIMERS

IN #1	5´-TAAGTTGGGTAACGCCAGGGTTTTC-3´
IN #2	5´-TTCCATGTTGCCACTCGCTTTAATG-3´
SEQ	5´-CCCCCTTAACGTGAGTTTTCGTTCC-3´

PROCEDURE

Day 1

Streak strain 10-1 for single colonies onto a YPD plate. Incubate for 2 days at 30°C.

Day 3

Pick an isolated colony and inoculate 5 ml of YPD liquid medium. Incubate overnight at 30°C.

Day 4

Determine the cell density using a hemocytometer (see Appendix G, Counting Yeast Cells with a Standard Hemocytometer Chamber). Dilute cells to 5 x 10^6 cells/ml and proceed with a high-efficiency transformation following Techniques and Protocols #1, High-efficiency Transformation of Yeast. There will be a slight modification as described below.

Harvest cells and transform with 9 μg of mTn3-*lacZ/LEU2* genomic library DNA previously digested with *Not*I. You will have a total of 9 μg of DNA in a total volume of 90 μl. Use 10 μl of DNA for each transformation. Perform ten transformations, one with no DNA as a control and nine with 1 μg of DNA. At step 16, resuspend the cells in 600 μl of sterile water and plate 200 μl of the transformants onto HC-leu plates. Incubate for 3 days at 30°C.

Strain 10-1 Transformation

Day 7

Determine the number of Leu$^+$ colonies from the strain 10-1 transformation. Lightly replica-plate the transformants to HC-leu and HC-leu containing 15 μg/ml benomyl. Incubate for 2 days at 23°C.

Day 9

Screen for benomyl sensitivity. Determine the proportion of benomyl-sensitive colonies among the transformants. Reclone six benomyl-sensitive colonies from HC-leu plates to new HC-leu plates. Incubate for 2 days at 30°C.

Day 11

Pick isolated colonies and grow 5-ml cultures in YPD overnight at 30°C.

Day 12

Frog cells to HC-Leu and HC-Leu+benomyl plates to retest the benomyl sensitivity. Incubate for 3 days at 23°C.

Day 13

Prepare genomic DNA from the six cultures using Techniques and Protocols #3D, Yeast Genomic DNA: Glass-bead Preparation, and resuspend the final pellet in 100 μl of TE. Digest 15 μl of each DNA with 5–10 U of *Rsa*I in a total volume of 100 μl. Add 1 μg of RNase A and incubate for several hours at 37°C. Inactivate the *Rsa*I by heating to 65°C for 20 minutes. Remove 10 μl and dilute to 100 μl in 1 x T4 DNA ligase buffer (with ATP). Add 100 U of T4 DNA ligase and incubate overnight at room temperature.

Day 14

Amplify the DNA adjacent to the Tn3 transposon by PCR. To 5 µl of ligated DNA, add

5 µl of 10x *Taq* polymerase buffer lacking Mg^{++}
1 µl of 25 µM primer 1
1 µl of 25 µM primer 2
1 µl of 10 mM dNTPs
32 µl of distilled H_2O
1 µl of *Taq* polymerase (2.5 U)

It is critical to use a "hot start" protocol for successful amplification of isolated genomic DNA. We use "Mg HotBeads" from Lumitekk. These are wax beads containing the $MgCl_2$ needed for the PCR reactions. Add 1 wax bead to each PCR reaction. The wax will melt at 68°C to release $MgCl_2$. Amplify for 5 minutes at 95°C, followed by 35 cycles of

1 minute at 94°C,
1 minute at 65°C,
2.5 minutes at 72°C.

Conclude the final cycle with 7 minutes at 72°C. Purify the amplified DNA away from protein, oligonucleotides, and dNTPs (a Wizard PCR preps kit is provided in the course), and submit for DNA sequencing using primer "SEQ."

Day 15

Score the benomyl sensitivity from plates that are at 23°C. Using a computer that is connected to the Internet, submit the DNA sequences for comparison to the yeast genome database and discover where your transposon has landed.

MATERIALS

Day 1 1 YPD plate

Day 3 1 culture tube containing 5 ml of YPD

Day 4 Erlenmeyer flask containing 50 ml of YPD
Materials for Techniques and Protocols #1, High-efficiency Transformation of Yeast
0.5 mg of carrier DNA
9 µg of *Not*I cleaved mTn3-*lacZ/LEU2* genomic library DNA
30 SC-leu plates

Day 7 30 SC-leu plates and 30 SC-leu plates containing 15 μg/ml of benomyl

Day 9 6 SC-leu plates

Day 11 6 culture tubes containing 5 ml of liquid YPD medium

Day 12 1 YPD plate
1 YPD+benomyl plate

Day 13 *Rsa*I
60 μl of 10x restriction enzyme buffer
5 μl of RNase A (5 mg/ml)
60 μl of T4 DNA ligase buffer+ATP
T4 DNA ligase

Day 14 Mg HotBeads
50 μl of 10x *Taq* polymerase buffer
10 μl of 25 M primer 1
10 μl of primer 2
10 μl of 10 mM dNTPs
dH_2O
Taq polymerase
Wizard PCR preps DNA purification system (Promega)
ABI Dye terminator DNA sequencing kit with mTn3 sequencing primer (Applied Biosystems Inc.)

Day 15 Networked computer

REFERENCES

Burns N., Grimwade B., Ross-Macdonald P.B., Choi E.-Y., Finberg K., Roeder G.S., and Snyder M. 1994. Large-scale characterization of gene expression, protein localization and gene disruption in *Saccharomyces cerevisiae*. *Genes Dev.* **8:** 1087–1105.

Ross-Macdonald P., Coelho P.S., Roemer T., Agarwal S., Kumar A., Jansen R., Cheung K.H., Sheehan A., Symoniatis D., Umansky L., Heidtman M., Nelson F.K., Iwasaki H., Hager K., Gerstein M., Miller P., Roeder G.S., and Snyder M. 1999. Large-scale analysis of the yeast genome by transposon tagging and gene disruption. *Nature* **402:** 413–418.

Isolation of Separation of Function Mutants by Two-hybrid Differential Interaction Screening

Most biological processes are mediated by protein-protein interactions, and a wealth of biochemical assays have been developed to detect such interactions. The two-hybrid system is a yeast-based genetic assay that provides a simple and sensitive means to detect potential interactions between two proteins. It is based on the finding that certain eukaryotic transcription activators are modular. For example, Gal4p, a positive regulator of galactose metabolism, consists of a site-specific DNA-binding domain (DBD) and an acidic transcription activation domain (Keegan et al. 1986). The DNA-binding domain binds to an upstream activating sequence (UAS) found in the promoters of *GAL* genes. The transcription activation domain interacts with other components of the transcription apparatus to initiate transcription. These two domains are separable and can function as independent units. Fields and Song (1989) used this feature of Gal4p to create a system in which two-hybrid proteins are created: One protein is fused to the Gal4p DNA-binding domain and another is fused to the Gal4p transcription activation domain. The binding and activation domains themselves do not interact with one another and, when expressed together in a yeast cell, are unable to activate transcription at a *GAL* promoter. However, if the sequences fused to the binding and activation domains are able to interact with one another, the binding and activation domains are brought together on the DNA and a functional transcriptional activator is reconstituted.

In the simplest case, the two-hybrid system is used to test for interactions between two known proteins; this is often referred to as a "directed" two-hybrid. A variation of the directed two-hybrid is to test the ability of mutant forms (typically deletions) of the proteins of interest to interact, potentially identifying domains required for the interaction. Another use of the two-hybrid system is to identify interacting proteins from a random genomic or cDNA library. In this case, the protein of interest is fused to the DNA-binding domain, and random genomic DNA or cDNA fragments are fused to the activation domain.

The activity of the hybrid transcriptional activator is assayed using constructs that have a *GAL* promoter driving one of a variety of reporter genes. In the version used in this experiment, the two reporter genes are *HIS3*, a yeast gene that encodes imidazoleglycerol-phosphate dehydratase, an enzyme involved in histidine biosynthesis, and *lacZ*, the bacterial gene encoding the enzyme β-galactosidase. The *HIS3* reporter allows selection of His^+ cells on medium lacking histidine. The *lacZ* reporter allows visual screening of cells expressing β-galactosidase, as visualized with the chromogenic substrate X-gal (5-bromo-4-chloro-3-indolyl-β-D-galactopyranoside).

Several requirements must be fulfilled for the two-hybrid method to succeed. First, both hybrid proteins must be expressed. In some versions of the two-hybrid vectors, an epitope tag is included in the construct, allowing expression to be verified by Western blotting. Second, neither hybrid protein should be able to activate transcription by itself. When identifying interactors from a library, there are several possibilities for false positives, not all of which are understood mechanistically. To help to identify such false positives, plasmids that are recovered from the library are subjected to additional tests: first, for self-activation of the reporter genes, then for specificity of interaction with the protein of interest, as compared to several standard test proteins (p53 and lamin, two proteins not found in yeast, are frequently used). As with all genetic methods, interactions detected in the two-hybrid system should be confirmed by biological and/or biochemical experiments.

As mentioned above, the two-hybrid system can be used to test if two proteins interact and to screen for interacting proteins from a library, but we have found that it can also be used to screen for mutants with specific protein interaction defects (see Fig. 1). This application requires that your protein of interest form interactions with multiple partners and that these interactions can be recapitulated in the two-hybrid system. In this experiment, we use inserts from two-hybrid fusions that have been amplified and randomly mutagenized with *Taq* polymerase (step 1). These polymerase chain reaction (PCR) products will be cotransformed with gapped two-hybrid vectors into two-hybrid tester strains (step 2). A recombination event between homologous sequences in the PCR product and the two-hybrid vector leads to gap repair, reforming the original two-hybrid fusion construct (Ma et al. 1987). The gap-repair event is selected by plating the transformants on medium that selects for the marker on the two-hybrid vector (*TRP1*). The transformant plates are then separately overlayed (by replica-plating) with two-hybrid tester strains of the opposite mating type that have been transformed with two different fusions to the reciprocal two-hybrid vector (step 3). The cells mate for one day, then the diploids that contain both two-hybrid vectors are selected. Two-hybrid activation is simultaneously assessed by the ability to grow in the presence of 3-amino-triazole, a competitive inhibitor of imidazoleglycerol-phosphate dehydratase (the His3p enzyme that is transcriptionally controlled by a *GAL*-responsive promoter). Mutants that fail to interact with one partner but are still able to interact with the other partner will be identified and purified. Here lies the strength of this approach: Highly disruptive mutants, that likely code for null alleles, are discarded whereas mutants with distinct defects (separation of function mutants) are readily isolated. Since our time is limited, you will be provided with both the gapped vector and the

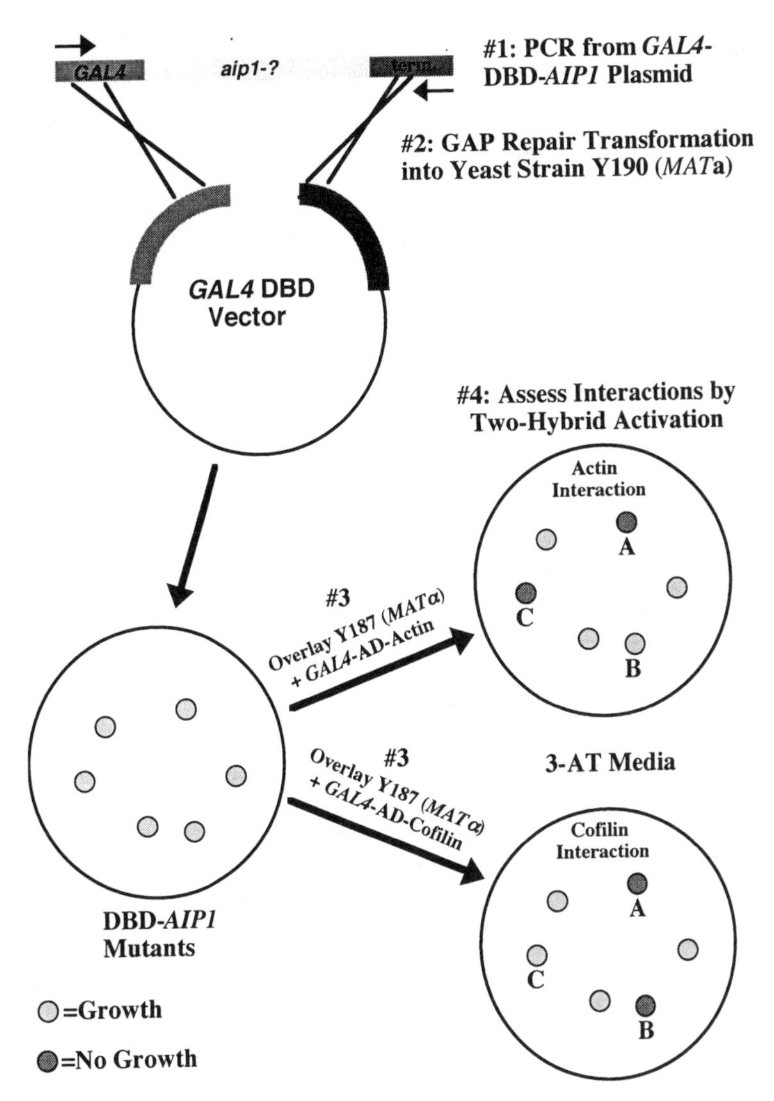

Figure 1. Using the two-hybrid system to screen for separation of function mutants in *AIP1*. In step 1, a mutagenic PCR is performed on an *AIP1* (actin interacting protein 1) insert in a *GAL4* DNA-binding domain (DBD) vector. In step 2, the PCR product is cotransformed with linearized/gapped *GAL4*-DBD vector DNA into yeast and the yeast reconstructs the original *GAL4* DBD-*AIP1* fusion construct by homologous recombination (gap repair). In step 3, yeast strains expressing fusions of the Gal4-activation domain to actin or cofilin are mated to the strains expressing Gal4-DBD fusions to the Aip1p mutants. The ability of the Aip1p mutants to interact with cofilin or actin is scored by activation of the His3p reporter as indicated by growth on medium containing the His3p inhibitor, 3-aminotriazole.

PCR mutagenized product and you will begin the experiment with the gap-repair transformation. The class will use a *GAL4*-based system developed by Steve Elledge (Durfee et al. 1993) to screen for mutants in *AIP1* (actin interacting protein 1) that are defective for either an interaction with actin or the actin binding protein cofilin (Rodal et al. 1999).

STRAINS

11-1	Y187	*MATα gal4 gal80 his3 trp1-901 ade2-101 ura3-52 leu2-3,112 cyh*^{r} *P_{GAL}-lacZ*
11-2	Y190	*MAT***a** *gal4 gal80 his3 trp1-901 ade2-101 ura3-52 leu2-3,112 URA3:: P_{GAL}-lacZ LYS2:: P_{GAL}-HIS3 cyh*^{r}
11-3	Y187	*MATα gal4 gal80 his3 trp1-901 ade2-101 ura3-52 leu2-3,112 cyh*^{r} *P_{GAL}-lacZ* [pAIP70]
11-4	Y187	*MATα gal4 gal80 his3 trp1-901 ade2-101 ura3-52 leu2-3,112 cyh*^{r} *P_{GAL}-lacZ* [pJT20]

PLASMIDS

pDAb189	Encodes a Gal4-DBD-Aip1p fusion in library screening vector pDAb1. This plasmid is the template for the mutagenic PCR and is a *TRP1*-marked, YCp vector.
pDAb1	A *GAL4*-DBD vector, *TRP1*-marked and YCp-based. This is the recipient plasmid for the gap-repair transformation.
pAIP70	Encodes a fusion of the Gal4 AD to actin in vector pACT, a *LEU2*-marked, YEp vector.
pJT20	Encodes a fusion of the Gal4 AD to cofilin in vector pACTII, a *LEU2*-marked, YEp vector.

PRIMERS

MCo-AIP1-DBD-1	5´-gactggaacagctatttctactg-3´ Sense strand primer to *GAL4* DBD in pDAb1, upstream of the polylinker.
MCo-AIP1-DBD-2	5´-gcagctggcacgacaggtttccc-3´ Antisense strand primer to the terminator in pDAb1, downstream from the polylinker.

PROCEDURE

Day BYA (before your arrival)

Mutagenic PCR

1 μl of pDAb189 (10 ng/λ of DNA)
5 μl of 5 μM MCo-AIP1-DBD-1
5 μl of 5 μM MCo-AIP1-DBD-2
5 μl of 10x *Taq* buffer
5 μl of 2 mM dNTPs

0.5 µl of BSA
1 µl of *Taq* polymerase
26.5 µl of H_2O

Run a standard PCR reaction
94°C x 4 minutes
25 cycles of
94°C x 1 minute
55°C x 1 minute
72°C x 2.5 minutes
72°C x 20 minutes
4°C soak

Vector preparation
Digest vector pDAb1 with *Sal*I and *Bam*HI for gap repair into yeast strain Y190
77.8 µl of H_2O
10 µg of pDAb1
10 µl of 10x *Bam*HI buffer
4 µl of *Eco*RI
4 µl of *Bam*HI
1 µl of 10 mg/ml BSA
Digest for 5 hours at 37°C
Extract the supernatant overnight at –20°C with an equal volume of 1:1 phenol-chloroform and ethanol precipitate by the addition of three volumes 100% ethanol

Next day
Centrifuge at 13,000 rpm for 10 minutes
Wash the pellet with 80% ethanol
Dry pellet and suspend in 80 µl of TE (pH 8.0)

Day 1

Start a 5-ml overnight of strain 11-2 (Y190) in YPD and incubate at 30°C in preparation for the gap-repair transformation the next morning.

Day 2

Inoculate 50 ml of YPD with the overnight culture of strain 11-2 (Y190) such that after at least three doublings, the final concentration will be 2×10^7 cells/ml. The doubling time of this strain is approximately 90 minutes, and one can assume there to be approximately 2×10^8 to 2.5×10^8 cells/ml in the overnight culture if it has reached stationary phase.

Follow Techniques and Protocols #1, High-efficiency Transformation of Yeast.

Perform a no-DNA control transformation, one transformation with 1 µl of cut pDAb1 DNA alone, and two transformations with 1 µl of cut pDAb1 plus 5 µl of *AIP1* PCR DNA. Reconstitute the cells into a final volume of 200 µl of sterile H_2O.

Plate all of the no-DNA control and pDAb1-alone transformations onto two SC-trp plates. Plate 20 and 180 µl of the gap-repair transformations onto two SC-trp plates (four plates total). Incubate at 30°C.

Day 3

Start 5-ml overnights of strains 11-3 (Y187+pAIP70) and 11-4 (Y187+pJT20) in SC-leu. Incubate at 30°C.

Day 4

Array 200–300 isolated colonies from the gap-repair transformation plates onto two to three fresh SC-trp plates.

Plate 200 µl of strains 11-3 (Y187+pAIP70) and 11-4 (Y187+pJT20) separately on two to three SC-leu plates to create a lawn. Incubate at 30°C.

Day 5

Replica-plate each plate of the arrayed gap-repair transformants onto two YPD plates. Onto one, replica-plate a lawn of strain 11-3 (Y187+pAIP70), and onto the other, replica-plate a lawn of strain 11-4 (Y187+pJT20). Be sure to mark the orientation of the plates. Incubate at 30°C. Save the original plates for the arrayed gap-repair transformants at 4°C.

Day 6

Replica-plate the mated cells to SC-trp-leu plates. Incubate at 30°C.

Day 8

Replica-plate the mated cells to SD+adenine+25 mM 3-AT, 50 mM 3-AT, and 100 mM 3-AT and incubate at 30°C.

Day 13

Identify the colonies that grew on the 3-AT plate with one mating partner but not the other. Go back to the original gap-repair transformation plate and recover the mutants of interest by streaking for single colonies onto fresh SC-trp plates.

We will stop the experiment at this stage but if we were to complete it, we would need to rescue the plasmids from yeast to *Escherichia coli* (Techniques and Protocols #3C, A Ten-minute DNA Preparation From Yeast; Release of Plasmid for Transformation of *E. coli* or Yeast). We would then retest the mutant by retransforming it into the two-hybrid tester strains. If it still failed to interact with only one binding partner, we would then have the mutant sequenced and tested for biological function.

REAGENTS AND MATERIALS

Day 1 5 ml of YPD

Day 2
50 ml of YPD
Materials for Techniques and Protocols #1, High-efficiency Transformation of Yeast
3 µl of *Bam*HI/*Sal*I cut pDAb1
10 µl of *AIP1* PCR
6 SC-trp plates

Day 3 10 ml of SC-leu liquid medium

Day 4
3 SC-trp plates
6 SC-leu plates

Day 5
6 YPD plates
12 sterile velvets

Day 6
6 SC-leu-trp plates
6 sterile velvets

Day 8
6 SD+adenine+25 mM 3-aminotriazole plates
6 SD+adenine+50 mM 3-aminotriazole plates
6 SD+adenine+100 mM 3-aminotriazole plates
6 sterile velvets

Day 13
2 SC-trp plates
Sterile toothpicks

REFERENCES

Durfee T., Becherer K., Chen P.-L., Yeh S.-H., Yang Y., Kilburn A.E., Lee W.-H., and Elledge S.J. 1993. The retinoblastoma protein associates with the protein phosphatase type 1 catalytic subunit. *Genes Dev.* **7:** 555–569.

Fields S. and Song O. 1989. A novel genetic system to detect protein-protein interactions. *Nature* **340:** 245–246.

Keegan L., Gill G., and Ptashne M. 1986. Separation of DNA binding from the transcription-activating function of a eukaryotic regulatory protein. *Science* **231:** 699–704.

Ma H., Kunes S., Schatz P.J., and Botstein D. 1987. Plasmid construction by homologous recombination in yeast. *Gene* **58:** 201–216.

Rodal A.A., Tetreault J.W., Lappalainen P., Drubin D.G., and Amberg D.C. 1999. Aip1p interacts with cofilin to disassemble actin filaments. *J. Cell Biol.* **145:** 1251–1264.

High-efficiency Transformation of Yeast

PROCEDURE

This protocol was adapted from Gietz and Schiestl (1995).

1. Inoculate 5 ml of liquid YPAD or 10 ml of SC and incubate with shaking overnight at 30°C.
2. Count overnight culture and inoculate 50 ml of YPAD to a cell density of 5 x 10^6/ml of culture.
3. Incubate the culture at 30°C on a shaker at 200 rpm until it is at 2 x 10^7 cells/ml. This typically takes 3–5 hours. This culture will give sufficient cells for ten transformations.

   Notes:
   i. It is important to allow the cells to complete at least two divisions.
   ii. Transformation efficiency remains constant for three to four cell divisions.
4. Harvest the culture in a sterile 50-ml centrifuge tube at 3000*g* (2500 rpm) for 5 minutes.
5. Pour off the medium, resuspend the cells in 25 ml of sterile H_2O, and centrifuge again.
6. Pour off the H_2O, resuspend the cells in 1.0 ml of 100 mM lithium acetate (LiAc), and transfer the suspension to a sterile 1.5-ml microfuge tube.
7. Pellet the cells at top speed for 5 seconds and remove the LiAc with a micropipette.
8. Resuspend the cells to a final volume of 500 µl (2 x 10^9 cells/ml), which is about 400 µl of 100 mM LiAc.

   Note: If the cell titer of the culture is greater than 2 x 10^7 cells/m, the volume of the LiAc should be increased to maintain the titer of this suspension at 2 x 10^9 cells/ml. If the titer of the culture is less than 2 x 10^7 cells/ml, decrease the amount of LiAc.
9. Boil a 1.0-ml sample of single-stranded carrier DNA for 5 minutes and quickly chill in ice water.

   Note: It is not necessary or desirable to boil the carrier DNA every time. Keep a small aliquot in your own freezer box and boil after three or four freeze/thaws. But keep on ice when out of box.
10. Vortex the cell suspension and pipette 50-µl samples into labeled microfuge tubes. Pellet the cells and remove the LiAc with a micropipette.

11. The basic "transformation mix" consists of the following ingredients; carefully add them *in the order listed:*

    240 µl of PEG (50% w/v)

    36 µl of 1.0 M LiAc

    25 µl of single-stranded carrier DNA (2.0 mg/ml)

    50 µl of H_2O and plasmid DNA (0.1–10 µg)

    Note: The order is important here. The PEG, which shields the cells from the detrimental effects of the high concentration of LiAc, should go in first.

12. Vortex each tube vigorously until the cell pellet has been completely mixed. This usually takes about 1 minute.
13. Incubate for 30 minutes at 30°C.
14. Heat shock for 20–25 minutes in a water bath at 42°C.

    Note: The optimum time can vary for different yeast strains. Test this if you need high efficiency from your transformations.

15. Microfuge at 6000–8000 rpm for 15 seconds and remove the transformation mix with a micropipette.
16. Pipette 0.2–1.0 ml of sterile H_2O into each tube and resuspend the pellet by pipetting it up and down gently.

    Note: Be as gentle as possible if high efficiency is important.

17. Plate from 200-µl aliquots of the transformation mix onto selective plates.

MATERIALS AND SOLUTIONS

Polyethylene glycol (PEG; 50% w/v) (MW 3350; Sigma P3640)

Make up to 50% (w/v) with H_2O and filter-sterilize with a 0.45-µm filter unit (Nalgene). Alternatively, the PEG solution can be autoclaved, but care must be taken to ensure that the PEG solution is at the proper concentration. In addition, it is important to store the PEG in a tightly capped container to prevent evaporation of H_2O and a subsequent increase in PEG concentration. Small variations above or below the PEG concentration optimum in the transformation reaction, which is 33% (w/v), can reduce the production of transformants.

Single-stranded carrier DNA (2 mg/ml)

High-molecular-weight DNA (deoxyribonucleic acid sodium salt type III from salmon testes; Sigma D1626)

TE buffer (pH 8.0)

10 mM Tris-HCl (pH 8.0)

1.0 mM EDTA

1. Weigh 200 mg of the DNA into 100 ml of TE buffer. Disperse the DNA into solution by drawing it up and down repeatedly in a 10-ml pipette. Mix vigorously on a magnetic stirrer for 2–3 hours or until fully dissolved. Alternatively, leave the covered solution, mixing at this stage overnight in a cold room.
2. Aliquot the DNA (100 µl is typically convenient) and store at –20°C.
3. Before using, an aliquot should be placed in a boiling water bath for at least 5 minutes and quickly cooled in an ice-water slurry.

Tips

i. Carrier DNA can be frozen after boiling and used three or four times. If transformation efficiencies begin to decrease with a batch of boiled carrier DNA, boil it again or use a new aliquot.

ii. The lower concentration of carrier DNA (2 mg/ml) in this protocol eases handling and gives more reproducible results.

iii. In previous protocol versions, a phenol:chloroform extraction was used to ensure maximal transformation efficiencies. This extraction may not be necessary if the DNA is of high enough quality. Test your carrier DNA to determine if extraction is necessary.

1.0 M lithium acetate (LiAc) stock solution

Prepare as a 1.0 M stock in distilled deionized H_2O; filter-sterilize. There is no need to titrate this solution, but the final pH should be between 8.4 and 8.9.

The latest version of this method can be found at
http://www.umanitoba.ca/faculties/medicine/human_genetics/gietz/method.html/.

REFERENCE

Gietz R.D. and Schiestl R.H. 1995. Transforming yeast with DNA. *Methods Mol. Cell. Biol.* **5:** 255–269.

"Quick and Dirty" Plasmid Transformation of Yeast Colonies

PROCEDURE

This protocol is modified from Chen et al. (1992).

1. Make this transformation mix fresh using a fresh stock of 2 M lithium acetate:

   200 µl of sterile 2 M lithium acetate
   800 µl of sterile 50% PEG-3350
   7.7 µl of β-mercaptoethanol

2. Add 3 µl of 10 mg/ml denatured salmon sperm DNA to a microfuge tube with 1 µg of plasmid DNA (5 µl from a standard miniprep is usually sufficient) plus 100 µl of transformation mix. Vortex to mix.
3. Suspend one large yeast colony (from a fresh streak) in the transformation plus DNA solution and vortex. Include a no-DNA control transformation.
4. Incubate for 30 minutes on a rotator at 37°C.
5. Pellet cells for 5 minutes at 3000 rpm, discard the supernatant, suspend the pellet in 100 µl of H_2O, and plate on selective medium.

MATERIALS AND SOLUTIONS

Transformation mix:
   50% polyethylene glycol (PEG) (MW 3350; Sigma P3640)
   2 M lithium acetate less than one month old, preferably fresh
   β-mercaptoethanol >98% (~17 M); Sigma M7154
   Carrier DNA (10 µg/µl of salmon or herring sperm)

Note: Cells from a fresh plate transform very efficiently; old cells that have been on plates for several months can also be transformed, but less efficiently.

REFERENCE

Chen D.C., Yang B.C., and Kuo T.T. 1992. One-step transformation of yeast in stationary phase. *Curr. Genet.* **21:** 83–84.

Yeast DNA Isolations

A. Yeast DNA Miniprep (40 ml)

PROCEDURE

1. Grow cells at 30°C to saturation (overnight) in 40 ml of YPD in a 125-ml flask.
2. Centrifuge the cells in a clinical centrifuge or a Sorvall SS-34 rotor at 5000 rpm for 5 minutes in a screw-capped centrifuge tube. Discard the supernatant.
3. Resuspend the cells in 3 ml of 0.9 M sorbitol, 0.1 M Na_2 EDTA (pH 7.5).
4. Add 0.1 ml of a 2.5 mg/ml solution of Zymolyase 100T and incubate for 1 hour at 37°C.
5. Centrifuge the cells in a clinical centrifuge or Sorvall SS-34 rotor for 5 minutes at 5000 rpm. Discard the supernatant.
6. Resuspend the cell pellet in 5 ml of 50 mM Tris-Cl (pH 7.4), 20 mM Na_2 EDTA.
7. Add 0.5 ml of 10% SDS and mix.
8. Incubate for 30 minutes at 65°C.
9. Add 1.5 ml of 5 M potassium acetate and store for 1 hour on ice.
10. Centrifuge in a Sorvall SS-34 rotor at 10,000 rpm for 10 minutes.
11. Transfer the supernatant to a fresh plastic centrifuge tube and add two volumes of 95% ethanol at room temperature. Mix and centrifuge at 5000–6000 rpm for 15 minutes at room temperature.
12. Discard the supernatant. Dry the pellet and then resuspend in 3 ml of TE (pH 7.4). This may take several hours.
13. Centrifuge in a Sorvall SS-34 rotor at 10,000 rpm for 15 minutes and transfer the supernatant to a new tube. Discard the pellet.
14. Add 150 µl of a 1 mg/ml solution of RNase A and incubate for 30 minutes at 37°C.
15. Add one volume of 100% isopropanol and shake gently to mix. Remove the precipitate, which should now look like a loose "cocoon" of fibers. Do not centrifuge. Air-dry.
16. Resuspend the precipitate in 0.5 ml of TE (pH 7.4). Store at 4°C. The final concentration of yeast DNA should be approximately 200 µg/ml. If the final solution is milky, reprecipitate the DNA with isopropanol or centrifuge in a Sorvall SS-34 rotor at 10,000 rpm for 15 minutes.

MATERIALS AND SOLUTIONS

125-ml flask containing 40 ml of YPD
Screw-capped centrifuge tubes
0.9 M sorbitol, 0.1 M Na_2 EDTA (pH 7.5) (3 ml)
Zymolyase 100T (120493-1, Seikagaku America Inc.) solution (0.1 ml)
2.5 mg/ml in 0.9 M sorbitol, 0.1 M Na_2 EDTA (pH 7.5)
50 mM Tris-Cl (pH 7.4), 20 mM Na_2 EDTA (5 ml)
10% SDS
5 M potassium acetate (1.5 ml)
95% ethanol
TE (pH 7.4)
    10 mM Tris-Cl (pH 7.4)
    1 mM Na_2 EDTA
RNase A solution (150 μl)
    Dissolve at a concentration of 1 mg/ml in 50 mM potassium acetate (pH 5.5). Boil for 10 minutes. Store frozen at –20°C.
100% isopropanol

B. Yeast DNA Miniprep (5 ml)

PROCEDURE

1. Grow cells overnight at 30°C in 5 ml of YPD.
2. Collect the cells in a clinical centrifuge at 2000 rpm for 5 minutes. Discard the supernatant.
3. Resuspend the cells in 0.5 ml of 1 M sorbitol, 0.1 M Na_2 EDTA (pH 7.5); transfer to a 1.5-ml microfuge tube.
4. Add 0.02 ml of a 2.5 mg/ml solution of Zymolyase 100T and incubate for 1 hour at 37°C.
5. Centrifuge in a microfuge for 1 minute.
6. Discard the supernatant. Resuspend the cells in 0.5 ml of 50 mM Tris-Cl (pH 7.4), 20 mM Na_2 EDTA.
7. Add 0.05 ml of 10% SDS; mix well.
8. Incubate the mixture for 30 minutes at 65°C.
9. Add 0.2 ml of 5 M potassium acetate and place the microfuge tube on ice for 1 hour.
10. Centrifuge in microfuge for 5 minutes.
11. Transfer the supernatant to a fresh microfuge tube and add one volume of 100% isopropanol at room temperature. Mix and allow it to sit for 5 minutes at room

temperature. Centrifuge *very briefly* (10 sec) in a microfuge. Pour off the supernatant and air-dry the pellet.

12. Resuspend the pellet in 0.3 ml of TE (pH 7.4).
13. Add 15 µl of a 1 mg/ml solution of RNase A and incubate for 30 minutes at 37°C (this step is optional).
14. Add 0.03 ml of 3 M sodium acetate and mix. Precipitate with 0.2 ml of 100% isopropanol. Centrifuge briefly again to collect the pellet of DNA.
15. Pour off the supernatant and air-dry. Resuspend the pellet of DNA in 0.1–0.3 ml of TE (pH 7.4).
16. Before using the DNA solution in a restriction digest, it may be necessary to centrifuge the final solution hard (15 minutes) in a microfuge to remove insoluble material that may inhibit digestion.

MATERIALS AND SOLUTIONS

YPD
1 M sorbitol, 0.1 M Na_2 EDTA (pH 7.5) (0.5 ml)
Zymolyase 100T (120493-1, Seikagaku America Inc.) solution (0.02 ml)
2.5 mg/ml in 1 M sorbitol, 0.1 M Na_2 EDTA (pH 7.5)
50 mM Tris-Cl (pH 7.4), 20 mM Na_2 EDTA (0.5 ml)
10% SDS
5 M potassium acetate (0.2 ml)
100% isopropanol
TE (pH 7.4)
  10 mM Tris-Cl (pH 7.4)
  1 mM Na_2 EDTA
RNase A solution (15 µl) (optional)
  Dissolve at a concentration of 1 mg/ml in 50 mM potassium acetate (pH 5.5). Boil for 10 minutes. Store frozen at –20°C.
3 M sodium acetate (0.03 ml)

C. A Ten-minute DNA Preparation from Yeast

Modified from Hoffman and Winston (1987).

SAFETY NOTES

Phenol is highly corrosive and can cause severe burns. It may be harmful by inhalation, ingestion, or skin absorption. Wear appropriate gloves, protective clothing, and safety glasses. All manipulations should be carried out in a chemical fume hood. Any areas of skin that come in contact with phenol should be rinsed with a large volume of water or polyethylene glycol 400 and washed with soap and water; ethanol should not be used.

Chloroform is irritating to the skin, eyes, mucous membranes, and upper respiratory tract. It is very volatile and should only be used in a chemical fume hood. Appropriate gloves and safety goggles should also be worn. Chloroform is a carcinogen and may damage the liver and kidneys.

Nitric acid is volatile and should be used in a chemical fume hood. It is toxic by inhalation, ingestion, and skin absorption. Concentrated acids should be handled with great care; appropriate gloves and a face protector should be worn. Keep away from heat, sparks, and open flame.

PROCEDURE

Release of Plasmid for Transformation of *E. coli* or Yeast

1. Grow small cultures (at least 1.4 ml) overnight at 30°C in a medium that maintains selection for the plasmid DNA, such as SC-ura.
2. Fill a 1.5-ml microfuge tube with the culture and collect the cells by a 5-second centrifugation in a microfuge.
3. Decant the supernatant and briefly vortex the tube to resuspend the pellet in the residual liquid.
4. Add 0.2 ml of 2% Triton X-100, 1% SDS, 100 mM NaCl, 10 mM Tris-Cl (pH 8), 1 mM Na_2 EDTA. Add 0.2 ml of phenol:chloroform:isoamyl alcohol (25:24:1). Add 0.3 g of acid-washed glass beads.
5. Vortex for 2 minutes.
6. Centrifuge for 5 minutes in a microfuge.
7. Transform 0.2 ml of competent *E. coli* cells with 1–5 μl of the aqueous layer. Transform yeast with 15 μl of the aqueous phase.

Isolation of Genomic DNA for Southern Blot Analysis

1. Grow 10-ml yeast cultures to saturation in YPD at 30°C.
2. Collect the cells by centrifugation for 2 minutes in a clinical centrifuge. Remove the supernatant and resuspend the cells in 0.5 ml of distilled H_2O. Transfer the cells to a 1.5-ml microfuge tube and collect them by centrifugation for 5 seconds in a microfuge.
3. Follow step 3 above.
4. Follow step 4 above.
5. Vortex for 3–4 minutes. Add 0.2 ml of TE (pH 8).
6. Centrifuge for 5 minutes in a microfuge. Transfer the aqueous layer to a fresh tube. Add 1 ml of 100% ethanol. Mix by inversion.
7. Centrifuge for 2 minutes in a microfuge. Discard the supernatant. Resuspend the pellet in 0.4 ml of TE plus 3 μl of a 10 mg/ml solution of RNase A. Incubate for 5

minutes at 37°C. Add 10 µl of 4 M ammonium acetate plus 1 ml of 100% ethanol. Mix by inversion.

8. Centrifuge for 2 minutes in a microfuge. Discard the supernatant. Air-dry the pellet and resuspend in 50 µl of TE. Use 10 µl for each sample to be analyzed by Southern blotting. This is ~2–4 µg of DNA.

MATERIALS AND SOLUTIONS

Medium that maintains selection for the plasmid DNA, such as SC-ura

2% Triton X-100, 1% SDS, 100 mM NaCl, 10 mM Tris-Cl (pH 8), 1 mM Na_2 EDTA (0.4 ml)

Phenol:chloroform:isoamyl alcohol (25:24:1) (0.4 ml)

Glass beads

0.45- to 0.5-mm beads are available from a variety of suppliers (e.g., Sigma, American QUALEX, Midwest Scientific, Stratagene). Beads can be cleaned by soaking in nitric acid and washing in copious amounts of distilled H_2O. Dry beads before use.

YPD

Sterile distilled H_2O

TE (pH 8)

10 mM Tris-Cl (pH 8)

1 mM Na_2 EDTA

100% ethanol

RNase A stock solution (3 µl)

Dissolve at a concentration of 10 mg/ml in 50 mM potassium acetate (pH 5.5). Boil for 10 minutes. Store frozen at –20°C.

4 M ammonium acetate (10 µl)

D. Yeast Genomic DNA: Glass-bead Preparation

1. Grow cells overnight in 5 ml of medium (rich or selective) at 30°C in a roller drum.
2. Transfer cultures to 13 x 100-mm glass tubes, and spin down cells in a tabletop centrifuge at 1500 rpm for 5 minutes.
3. Wash cells with 3 ml of sterile H_2O and spin as above.
4. Resuspend in 500 µl of lysis buffer.
5. Add clean glass beads (about two thirds of a 1.5-ml Eppendorf tube) and 25 µl of 5 M NaCl.
6. Vortex on highest setting for 1 minute.
7. Spin at 2000 rpm for 2 minutes.
8. Transfer the liquid with a P-1000 to a 1.5-ml Eppendorf tube.
9. Add 500 µl of phenol, vortex, and spin for 1 minute. Extract aqueous layer (top) with a P-1000 and transfer to a clean tube. Add 500 µl of SEVAG (24:1 chloroform:isoamyl alcohol), vortex, spin, and extract as above.

10. Add 1 ml of cold 95% ethanol and precipitate for 1 hour at –20°C.
11. Pellet the DNA by spinning for 5 minutes at full speed, pour off the supernatant, and wash with 70% ethanol. Resuspend in 250 µl of TE.
12. Add 25 µl of EDTA-Sark and 5 µl of proteinase K (10 mg/ml). Incubate for 30 minutes at 37°C.
13. Add 250 µl of 5 M NH_4Ac and repeat steps 9 and 10.
14. Pellet the DNA, wash with 70% ethanol, and resuspend in 100 µl of TE (use ~10 µl/digest).

MATERIALS AND SOLUTIONS

YPD
13 x 100-mm glass tubes
Sterile H_2O
Lysis buffer
- 0.1 M Tris-Cl (pH 8.0)
- 50 mM EDTA
- 1% SDS

5 M NaCl
Glass beads (0.5 mm in diameter; BioSpec Products 11079-105)
SEVAG (24:1 chloroform:isoamyl alcohol)
95% ethanol
70% ethanol
EDTA-Sark
- 0.4 M EDTA (pH 8.0)
- 2% *N*-lauroylsarcosine (Sarkosyl)

Proteinase K (10 mg/ml)
5 M $NH_4(C_2H_3O_2)$
TE
- 10 mM Tris-Cl
- 1 mM EDTA (pH 8.0)

Phenol, equilibrated with H_2O (see Safety Notes in part C above)

REFERENCE

Hoffman C.S. and Winston F. 1987. A ten-minute DNA preparation from yeast efficiently releases autonomous plasmids for transformation of *Escherichia coli*. *Gene* **57**: 267–272.

Yeast Protein Extracts

This protocol is a simple, reliable method for preparation of yeast protein extracts for PAGE analysis and Western blotting. This procedure works for both growing and stationary phase cells, grown either in liquid or on plates (YPD or minimal). A sample of 2.3 mg (wet weight) of cells yields sufficient protein to run several lanes on a minigel. The technique was adapted from Kushnirov (2000).

PROCEDURE

1. Collect about 2.5 OD_{600} of cells (approximately 2.3 mg wet weight) from a liquid culture or scraped from the surface of a plate with a bacteriological loop. Resuspend the cells in 100 μl of distilled water, add 100 μl of 0.2 M NaOH, and incubate the sample for 5 minutes at room temperature.
2. Pellet the sample, and resuspend it in 50 μl of PAGE sample buffer. Boil the sample for 3 minutes and pellet again. Use a pipette to remove the supernatant, which contains protein. Use 6 μl of the supernatant per lane and resolve on a PAGE minigel.
3. To measure protein concentration by Bradford assay, dilute the sample 1:3000.

MATERIALS AND SOLUTIONS

0.2 M sodium hydroxide solution
2x PAGE sample buffer
120 mM Tris/HCl (pH 6.8)
10% glycerol
4% SDS
8% β-mercaptoethanol (I use 5%)
0.004% bromophenol blue

REFERENCE

Kushnirov V.V. 2000. Rapid and reliable protein extraction from yeast. *Yeast* **16:** 857–860.

Techniques and Protocols #5

TAP Purification

PROCEDURE

1. Grow 3–6 liters of cells to 2 x 10^7 to 3 x 10^7 cells per ml (OD = 1.0–1.3).
2. Pellet cells using centrifuge bottles and wash once with ice cold water.
3. Wash once with ice cold NP-40 buffer while transferring to 50-ml centrifuge tubes.
4. Pellet cells and resuspend the cells in 10 ml of NP-40 buffer.
5. Prechill a mortar and pestle with dry ice. Add liquid nitrogen and allow to completely cool. When the liquid nitrogen evaporates, add cells dropwise to the pestle to freeze. Occasionally add a little liquid nitrogen to keep the cells cool. Grind with the pestle, very hard, for 15 minutes. Do not allow cells to thaw.
6. Transfer lysate to an ice cold beaker and thaw at room temperature. When the edges are thawed, add 20 ml of NP-40 buffer containing protease inhibitors (there are many commercially available cocktails).
7. Transfer crude lysate to 40-ml Nalgene tubes and spin at 16,500 rpm in a SA-600 rotor or equivalent.
8. Add 1 ml of Sepharose 6B (Sigma) beads (1:1 slurry with NP-40 buffer). Incubate on a rotating platform for 30 minutes at 4°C. Pellet the beads. Transfer the supernatant into a new 50-ml Falcon tube.
9. Add 400 µl of IgG Sepharose 6 Fast Flow (Amersham) (previously prepared in a 1:1 slurry with NP-40 buffer) and incubate on a rotating platform for at least 2 hours at 4°C.
10. Pour the lysate/IgG Sepharose suspension onto a BioRad Poly-Prep chromatography column with a reservoir. Pack by gravity.
11. Wash beads with 30 ml of NP-40 buffer.
12. Wash beads with 10 ml of TEV C-buffer.
13. Close the bottom with a stopper and add 1 ml of TEV C-buffer and 5 µg of TEV. Plug the top of the column and incubate on a rotating platform between 2 hours and overnight at 4°C.
14. Drain eluate into a new column sealed at the bottom.

15. Wash out old column with 1.0 ml of TEV C-buffer.
16. Add 3 ml of CBB buffer to the TEV supernatant plus 3 µl of 1 M $CaCl_2$ per ml of TEV eluate. Add 300 µl of calmodulin Sepharose (Amersham) in CBB buffer (1:1 slurry) and incubate on the rotating platform for 1 hour at 4°C.
17. Wash:
    a. Twice in 1 ml of CBB (0.1% NP-40).
    b. Once in 1 ml of CBB (0.02% NP-40).
18. Plug the bottom of the column and add 1 ml of CEB.
19. Elute first 1-ml fraction into a microfuge tube.
20. Plug the bottom of the column and add 1 ml of CEB.
21. Elute second 1-ml fraction into a microfuge tube.
22. Combine the fractions and split into 500-, 750-, and 750-µl portions. Place in non-siliconized microfuge tubes.
23. Adjust aliquots to 25% TCA with 100% TCA and place on ice for 30 minutes with periodic vortexing.
24. Spin at maximum speed (13,000 rpm) in a microfuge for 30 minutes at 4°C.
25. Wash once with ice cold (–20°C) acetone containing 0.05 N HCl and spin for 5 minutes at maximum speed (13,000 rpm) at 4°C.
26. Wash once with ice cold (–20°C) acetone and spin for 5 minutes at maximum speed at 4°C.
27. Remove supernatant and dry in a speed vacuum for approximately 10 minutes.

Use the 500-µl fraction for silver staining on a 10% gel. Use 750-µl fractions for mass spectrometry.

NP-40 Buffer

10 mM sodium phosphate buffer (pH 7.2)
150 mM sodium chloride
1% NP-40
50 mM sodium fluoride
0.1 mM sodium vanadate
1 mM DTT
Protease inhibitors

TEV C-buffer

25 mM Tris (pH 8.0)
150 mM sodium chloride

0.1% NP-40
0.5 mM EDTA
1 mM DTT

Calmodulin Binding Buffer (CBB)

25 mM Tris (pH 8.0)
150 mM sodium chloride
1 mM magnesium acetate
1 mM imidazole
2 mM calcium chloride
10 mM β-mercaptoethanol

Calmodulin Elution Buffer (CEB)

25 mM Tris (pH 8.0)
150 mM sodium chloride
1 mM magnesium acetate
1 mM imidazole
0.02% NP-40
20 mM EGTA
10 mM β-mercaptoethanol

Yeast RNA Isolations

LARGE-SCALE RNA ISOLATION PROCEDURE

This procedure is designed to yield more than 10 mg of total nucleic acid and 300–400 µg of poly(A)-selected mRNA from each 500-ml culture. The volumes indicated can be applied directly to cultures of 200–500 ml but should be adjusted for cultures of significantly smaller or larger volumes.

It is convenient to inoculate 200 ml of YPD (or SC) with an appropriate volume of a fresh overnight culture to yield a concentration of 2×10^7 to 4×10^7 cells/ml the following morning. The cell density should be checked using a hemocytometer (see Appendix G, Counting Yeast Cells with a Standard Hemocytometer Chamber) or Klett (50–100 Klett units for most strains).

PROCEDURE

SAFETY NOTES

Nitric acid is volatile and should be used in a chemical fume hood. Concentrated acids should be handled with great care; gloves and a face protector should be worn.

Phenol is highly corrosive and can cause severe burns. Gloves, protective clothing, and safety glasses should be worn. All manipulations should be carried out in a chemical fume hood. Any areas of skin that come in contact with phenol should be rinsed with a large volume of water or polyethylene glycol 400 and washed with soap and water; ethanol should not be used.

Diethyl pyrocarbonate (DEPC) is toxic and volatile. Work with open tubes in a chemical fume hood and wear gloves. Furthermore, addition of DEPC to Tris-HCl can lead to combustion on heating.

Chloroform is irritating to the skin, eyes, mucous membranes, and upper respiratory tract. It should be used only in a chemical fume hood. Gloves and safety goggles should also be worn. Chloroform is a carcinogen and may damage the liver and kidneys.

Isolation of Total Yeast RNA

1. Add 50 µg of cycloheximide for each ml of culture. Shake for 15 minutes at 30°C. This step is optional but is supposed to protect mRNA by freezing it in polysomes.
2. To quickly cool the cells for harvesting, fill large centrifuge bottles (Sorvall GS3 or equivalent) halfway with crushed ice, pour the culture over the ice, and shake. Centrifuge at 5000 rpm for 5 minutes at 4°C.
3. Add 11 g of acid-washed glass beads to a 30-ml centrifuge tube. Corex glass tubes work best, but plastic tubes can be used.
4. Add 3 ml of phenol equilibrated with LETS buffer to the glass beads.
5. Resuspend the cell pellet in 2.5 ml of ice cold LETS buffer and add it to the glass beads/phenol. The meniscus should be just above the surface of the glass beads for the best cell breakage.
6. Vortex at top speed, alternating 30 seconds of vortexing with 30 seconds on ice, for a total of 3 minutes of vortexing. Cell breakage can be checked with a phase-contrast microscope. Broken cells appear as nonrefractile "ghosts."
7. When at least 90% of the cells are broken, add 5 ml of ice cold LETS buffer and vortex briefly. Centrifuge for 5 minutes at 8000 rpm in a Sorvall SS-34 rotor to break the phases. After centrifugation, the phenol layer plus interface should be just below the surface of the beads, allowing easy retrieval of the aqueous phase without disturbing the interface.
8. Transfer the aqueous phase to a clean tube and extract twice with 5 ml of phenol:chloroform:isoamyl alcohol (25:24:1). Avoid transferring the interface. Extract once with chloroform (optional).
9. Add one-tenth volume of 5 M LiCl and precipitate for 3 hours at –20°C. The RNA precipitate may be conveniently stored at –20°C at this point.

Isolation of Poly(A)$^+$ RNA

1. Centrifuge at 10,000 rpm for 10 minutes in a Sorvall SS-34 rotor to pellet the stored RNA. Wash with 80% ethanol. Dry under a vacuum and dissolve in 2.5 ml of distilled H_2O. When dissolved, add 2.5 ml of 2x loading buffer and add SDS to a final concentration of 0.3%.
2. Heat for 5 minutes at 65°C.
3. Load an oligo(dT) column and wash three times with 1x loading buffer.
4. Elute the poly(A)$^+$ RNA with 10 mM HEPES (pH 7); add RNase inhibitor if desired.
5. Determine the RNA concentration by measuring the OD_{260} of a 1:10 dilution of the poly(A)$^+$ RNA in distilled H_2O. Compare this with a 1:10 dilution of 10 mM HEPES (pH 7) in H_2O (i.e., compare with 1 mM HEPES).
6. Precipitate the RNA by adding one-tenth volume of 3 M sodium acetate and 2.5 volumes of 100% ethanol. Mix by inversion and incubate for 30 minutes at –70°C.

Centrifuge for 10 minutes in a microfuge. Remove the supernatant and dissolve the RNA pellet in distilled H_2O to make a final concentration of 1 mg/ml. Both the dissolved RNA and the ethanol precipitate can be stored indefinitely at –70°C.

MATERIALS

Cycloheximide (50 µg/ml of culture) (optional)
Large centrifuge bottles (Sorvall GS3 or equivalent)
Glass beads
   0.45- to 0.5-mm beads are available from a variety of suppliers (e.g., Sigma, American QUALEX, Midwest Scientific, Stratagene). Beads can be cleaned by soaking in nitric acid and washing in copious amounts of distilled H_2O. Dry beads before use.
Centrifuge tubes (Corex glass or plastic; 30 ml)
Phenol equilibrated with LETS buffer (3 ml)
LETS buffer
   0.1 M lithium chloride (LiCl)
   0.01 M Na_2 EDTA
   0.01 M Tris-Cl (pH 7.4)
   0.2% SDS
   0.1% diethyl pyrocarbonate (optional)
Phenol:chloroform:isoamyl alcohol (25:24:1)
Chloroform (optional)
5 M LiCl
80% ethanol
Sterile distilled H_2O
1x loading buffer
   0.5 M NaCl
   0.01 M HEPES (pH 7)
SDS
Oligo(dT) column
10 mM HEPES (pH 7)
RNase inhibitor (optional)
3 M sodium acetate
100% ethanol

RAPID AND SMALL-SCALE RNA ISOLATION

This method is based on that of Schmitt et al. (1990).

1. Inoculate a 50-ml YPD culture with 7.5×10^7 cells from a confluent overnight culture.

2. Incubate cultures at appropriate temperature and agitation until desired concentration is achieved. Cells can be shifted to various temperatures before harvesting. Cell densities should not exceed 1.5 x 10^7 cells/ml or an OD_{600} of 1.5. Density determination with a hemocytometer (Appendix G, Counting Yeast Cells with a Standard Hemocytometer Chamber) may prove useful for later comparison of relative RNA expression levels.

3. Harvest the cells by centrifugation at 3000*g* for 3 minutes at 4°C. Decant liquid and resuspend cells with 1 ml of chilled AE buffer. Quickly transfer cells to microcentrifuge tubes and pellet cells by centrifugation at 16,000*g* for 10 seconds in a microcentrifuge. Decant and resuspend cells in 1 ml of chilled AE buffer. Repeat centrifugation and decanting steps.

   Important: Cells must be chilled as quickly as possible and kept cold throughout the entire assay to reduce RNA degradation.

4. Resuspend cells in 400 µl of chilled AE buffer. Add 40 µl of 10% SDS and mix thoroughly. Add 440 µl of phenol previously equilibrated to pH 5.2 with AE buffer and mix thoroughly.

5. Prepare a dry ice/95% ethanol bath: The dry ice should be crushed to powder and slowly added to the 95% ethanol until a thick slurry is achieved.

6. Flash freeze cells in the dry ice/ethanol bath for 5 minutes and then transfer to a 65°C water bath for 5 minutes. Vortex lysates for 30 seconds and repeat the freeze/thaw/vortex cycle.

7. Flash freeze cells again in the dry ice/ethanol bath for 5 minutes.

8. Pellet lysates using a microcentrifuge at 16,000*g* for 7 minutes at room temperature. If separation of organic and aqueous layers appears incomplete (large white fluffy suspended material), pellet lysates again at 16,000*g* for 5 minutes. If this fails to separate the layers, add another 100 µl of chilled AE buffer and 100 µl of phenol (pH 5.2) and pellet lysates again in a microcentrifuge at 16,000*g* for 5 minutes at room temperature. Transfer aqueous layer to a new microcentrifuge tube.

9. Add 600 µl of phenol (pH 5.2)-chloroform-isoamyl alcohol (25:24:1) to extract, vortex, and pellet extracts again in a microcentrifuge at 16,000*g* for 5 minutes at room temperature. Transfer aqueous layer to a new microcentrifuge tube.

10. The RNA is precipitated by adding 50 µl of 3 M sodium acetate (pH 5.2) and 1 ml of chilled 100% ethanol to extract. Chill RNA for 20 minutes to –20°C. RNA can be stored until needed at –70°C in 100% ethanol without sodium acetate.

11. Pellet the RNA at 16,000*g* in a microcentrifuge for 15 minutes at 4°C. Decant completely and wash RNA pellet with 1 ml of chilled 80% ethanol. Pellet the RNA at 16,000*g* in a microcentrifuge for 5 minutes and decant all liquid (removal of all liquid reduces salt contamination). Dry the RNA in a vacuum centrifuge (5–10 min) and then resuspend RNA in 100 µl of ddH_2O.

12. Determine absorbance at 260 and 280 nm for RNA concentration and purity analysis. A 260/280 nm ratio of 1.7–2.0 indicates RNA of acceptable purity. 1 abs. unit = 38 µg of RNA. RNA yield ranges from 50 to 200 µg.

    Note: The above assay should be done in containers washed in RNase Away™ (LPS) or dry baked for 24 hours at 160°C. All aqueous reagents should be made with 1:1000 volume of diethyl pyrocarbonate (DEPC) before autoclaving to inactivate RNases.

MATERIALS

AE buffer
- 50 mM sodium acetate (pH 5.2)
- 10 mM EDTA (pH 8.0)

10% SDS
Phenol equilibrated in AE buffer (pH 5.2)
Dry ice/95% ethanol bath
3 M sodium acetate (pH 5.2)
100% ethanol
80% ethanol
Diethyl pyrocarbonate (DEPC; Sigma)

REFERENCE

Schmitt M.E., Brown T.A., and Trumpower B.L. 1990. A rapid and simple method for preparation of RNA from *Saccharomyces cerevisiae*. *Nucleic Acids Res.* **18:** 3091–3092.

Hydroxylamine Mutagenesis of Plasmid DNA

PROCEDURE

SAFETY NOTES

Hydroxylamine is corrosive and toxic. It may be harmful by inhalation, ingestion, or skin absorption. Wear appropriate gloves and safety glasses and use only in a chemical fume hood.

1. Prepare the hydroxylamine solution just before use and store on ice until needed.
2. Add 10 µg of CsCl-purified plasmid DNA to 500 µl of hydroxylamine solution in a microfuge tube.
3. Incubate for 20 hours at 37°C.
4. Stop the reaction by adding 10 µl of 5 M NaCl, 50 µl of 1 mg/ml BSA, and 1 ml of 100% ethanol; precipitate the DNA for 10 minutes at –70°C.
5. Centrifuge the precipitated DNA in a microfuge for 10 minutes. Carefully remove all of the supernatant.
6. Resuspend the DNA in 100 µl of TE (pH 8). Add 10 µl of 3 M sodium acetate and 250 µl of 100% ethanol; precipitate the DNA for 10 minutes at –70°C and centrifuge as in step 5.
7. Allow the pellet to air-dry and then resuspend it in 100 µl of TE (pH 8).
8. The DNA can be used directly for transformation of either *E. coli* or yeast. The transformation frequency in yeast with the mutagenized DNA is reduced only approximately threefold relative to unmutagenized DNA. The formation of Ura^- plasmids in *E. coli* can be monitored. The *E. coli* strain can be ung^+, but the transformation frequency will be reduced 10- to 100-fold. When the *URA3* gene makes up approximately 10% of the plasmid DNA and an *E. coli* ung^+ $pyrF^-$ strain (DB6507) is being transformed, approximately 4% Ura^- colonies can be expected. Transforming yeast with the same stock of mutagenized DNA gives approximately 1% loss-of-function mutations ("knockouts") in my favorite gene. Approximately 10% of the mutants are temperature sensitive.

MATERIALS AND SOLUTIONS

Hydroxylamine solution

- 0.35 g of hydroxylamine HCl
- 0.09 g of NaOH
- 5 ml of distilled H_2O (ice cold)

  Dissolve the solids in the H_2O. The pH should be approximately 7. Prepare just before use and store on ice until needed.

CsCl-purified plasmid DNA (10 µg)

5 M NaCl

1 mg/ml bovine serum albumin (BSA)

100% ethanol

TE (pH 8)

- 10 mM Tris-Cl (pH 8)
- 1 mM Na_2 EDTA

3 M sodium acetate

Assay of β-Galactosidase in Yeast

There are two basic methods for the in vitro assay of β-galactosidase in yeast. They differ mainly in the method of preparing the material for assay. In the first method (Rose and Botstein 1983), a crude extract is prepared, and the activity is normalized to the amount of protein assayed. In the second method (Guarente 1983), the cells are permeabilized to allow the substrate to enter the cells and the activity is normalized to the number of cells assayed. The former method is preferable when comparing cells that are grown under very different conditions or that have different genetic backgrounds. The latter method is adapted from the assay for *E. coli* and is particularly suited for changing levels of activity within a single strain. The third method lyses the cells by a simple freeze/thaw protocol and utilizes the chemiluminescent assay to measure β-galactosidase levels.

METHOD 1: ASSAY OF CRUDE EXTRACTS

SAFETY NOTES

Nitric acid is volatile and must be handled with great care. It is toxic by inhalation, ingestion, and skin absorption. Wear appropriate gloves and safety goggles and use in a chemical fume hood. Do not breathe the vapors. Keep away from heat, sparks, and open flame.

Phenylmethylsulfonyl fluoride (PMSF) is a highly toxic cholinesterase inhibitor. It is extremely destructive to the mucous membranes of the respiratory tract, eyes, and skin. It may be fatal by inhalation, ingestion, or skin absorption. Wear appropriate gloves and safety glasses and always use in a chemical fume hood. In case of contact, immediately flush eyes or skin with copious amounts of water and discard contaminated clothing.

Chloroform is irritating to the skin, eyes, mucous membranes, and respiratory tract. It is a carcinogen and may damage the liver and kidneys. It is also volatile. Avoid breathing the vapors. Wear appropriate gloves and safety glasses and always use in a chemical fume hood.

1. Grow a 5-ml culture of cells to a concentration of 1×10^7 to 2×10^7 cells/ml in an appropriate liquid medium at an appropriate temperature (usually 30°C). If the hybrid gene is expressed from an autonomous plasmid, use an appropriate medium to select for the presence of the plasmid.

2. Chill the cells on ice and harvest by centrifugation (2000 rpm for 5 minutes in a clinical centrifuge is adequate). Keep the cells on ice from this point on.
3. Discard the supernatant. Resuspend the cells in 250 µl of breaking buffer. The cells can now be frozen at –20°C and assayed at a later date. All of the following steps can be performed in a 1.5-ml microfuge tube.
4. If the cells were frozen, thaw them on ice. Add glass beads until the beads reach a level just below the meniscus of the liquid. Add 12.5 ml of PMSF stock solution.
5. Vortex six times at top speed in 15-second bursts. Chill on ice between bursts.
6. Add 250 ml of breaking buffer and mix well. Withdraw the liquid extract after plunging the tip of a 1000-ml pipettor to the bottom of the tube.
7. Clarify the extract by centrifuging for 15 minutes in a microfuge. If the activity is in the particulate fraction, the unclarified supernatant can be used and the assay mixture may be clarified later in step 8.
8. To perform the assay,
   a. Add 10–100 µl of extract to 0.9 ml of Z buffer. Adjust the volume to 1 ml with breaking buffer.
   b. Incubate the mixture in a water bath for 5 minutes at 28°C.
   c. Initiate the reaction by adding 0.2 ml of ONPG stock solution. Note precisely the time that the addition is made. Incubate at 28°C until the mixture has developed a pale yellow color.
   d. Terminate the reaction by adding 0.5 ml of Na_2CO_3 stock solution. Note precisely the time that the reaction is terminated. Measure the optical density at 420 nm.
9. Measure the protein concentration in the extract using the dye-binding assay of Bradford (1976).
   a. Dilute the Bradford reagent fivefold in distilled H_2O. Filter the diluted reagent through Whatman 540 paper (or equivalent). Typical extracts prepared in this fashion contain 0.5–1 mg/ml of protein.
   b. Add 10–20 µl of the extract to 1 ml of the diluted reagent and mix. Measure the blue color formed at 595 nm. Use disposable plastic cuvettes to prevent the formation of a blue film.
   c. Prepare a standard curve using several dilutions (0.1–1 mg/ml) of BSA dissolved in breaking buffer.
10. Express the specific activity of the extract according to the following formula:

$$\frac{OD_{420} \times 1.7}{0.0045 \times \text{protein} \times \text{extract volume} \times \text{time}}$$

OD_{420} is the optical density of the product, *o*-nitrophenol, at 420 nm. The factor 1.7 corrects for the reaction volume. The factor 0.0045 is the optical density of a 1 nmole/ml solution of *o*-nitrophenol. Protein concentration is expressed as mg/ml. Extract volume is the volume assayed in ml. Time is in minutes. Specific activity is expressed as nmoles/minute/mg of protein.

METHOD II: PERMEABILIZED CELL ASSAY

1. Grow the cells as above. Measure the OD_{600} of the culture and harvest 1×10^6 to 1×10^7 cells by centrifugation as above.
2. Discard the supernatant. Resuspend the cells in 1 ml of Z buffer.
3. Add 3 drops of chloroform and 2 drops of 0.1% SDS. Vortex at top speed for 10 seconds.
4. Preincubate the samples for 5 minutes at 28°C. Start the reaction by adding 0.2 ml of ONPG as above.
5. Stop the reaction by adding 0.5 ml of Na_2CO_3 stock solution when the sample in the tube has developed a pale yellow color. Note the amount of time that has elapsed during the assay. Remove the cell debris by centrifuging for 10 minutes in a microfuge and then discard the pellet.
6. Measure the OD_{420} of the reactions.
7. Express the activity as β-galactosidase units:

$$\frac{OD_{420}}{OD_{600} \text{ of assayed culture} \times \text{volume assayed} \times \text{time}}$$

OD_{420} is the optical density of the product, *o*-nitrophenol. OD_{600} is the optical density of the culture at the time of assay. Volume is the amount of the culture used in the assay in ml. Time is in minutes.

MATERIALS AND SOLUTIONS

Appropriate liquid medium

Breaking buffer

- 100 mM Tris-Cl (pH 8)
- 1 mM dithiothreitol
- 20% glycerol

Glass beads

0.45- to 0.5-mm beads are available from a variety of suppliers (e.g., Sigma, American QUALEX, Midwest Scientific, Stratagene). Beads can be cleaned by soaking them in nitric acid and washing them in copious amounts of distilled H_2O. Dry beads before use.

PMSF (Sigma P7626) stock solution

40 mM in 100% isopropanol. Store at –20°C.

Z buffer (Miller 1972)

16.1 g of $Na_2HPO_4 \cdot 7H_2O$

5.5 g of $NaH_2PO_4 \cdot H_2O$

0.75 g of KCl

0.246 g of $MgSO_4 \cdot 7H_2O$

2.7 ml of β-mercaptoethanol

Distilled H_2O to make a final volume of 1 liter

Adjust the pH to 7. Store at 4°C.

ONPG (*o*-nitrophenyl-β-D-galactoside) stock solution

4 mg/ml in Z buffer. Store at –20°C.

Na_2CO_3 stock solution

1 M in distilled H_2O

Bradford reagent (Bio-Rad)

Distilled H_2O

Whatman 540 paper or equivalent

Disposable plastic cuvettes

0.1–1 mg/ml bovine serum albumin (BSA) in breaking buffer

Chloroform

0.1% SDS

METHOD III: FREEZE/THAW ASSAY FOR β-GALACTOSIDASE BY CHEMILUMINESCENCE

These procedures use a chemiluminescent substrate for β-galactosidase that is marketed by Applied Biosystems for Tropix. The cells must be lysed to allow the substrate and β-galactosidase to meet. These freeze/thaw protocols are simple and are also compatible with luciferase assays.

PROCEDURE FOR MICROFUGE TUBES

1. Grow cells overnight to saturation (about 10^8 cells per ml).
2. Dilute cells 50-fold and grow for 6 hours.
3. Transfer 1 ml to microfuge tube and pellet. Pour off all liquid.
4. Freeze in liquid nitrogen for 2 minutes.
5. Thaw for 2 minutes in room-temperature water.
6. Repeat freeze/thaw cycle.
7. Resuspend cells in 100 μl of Z buffer from Galacto-Star system by Tropix.

8. Add 20 μl of lysed cells to 100 μl of β-galactosidase substrate and buffer from Tropix.
9. Wait 60 minutes and measure sample in luminometer.

PROCEDURE FOR 96-WELL PLATES

1. Grow cells overnight to saturation in 0.6 ml in deep well (2 ml) 96-well plate.
2. Dilute cells 50-fold into standard 96-well plates and grow for 6 hours.
3. Centrifuge plate and invert to remove liquid. Blot on paper towels.
4. Place for 1 hour at –80°C (stable for at least 2 days).
5. Thaw plate and add 100 μl of Z buffer from Galacto-Star system by Tropix.
6. Add 100 μl of β-galactosidase substrate and buffer from Tropix to appropriate number of wells of a white, opaque, 96-well plate.
7. Transfer 20 μl of cell extract wells containing substrate to a white assay plate.
8. Wait 60 minutes and assay in 96-well luminometer.

MATERIALS AND SOLUTIONS

Applied Biosystems markets a variety of chemiluminescence kits to assay β-galactosidase. The Tropix Gal-Screen System (catalog numbers GSY200 and GSY1000) works on yeast.

REFERENCES

Bradford M.M. 1976. A rapid and sensitive method for the quantitation of microgram quantities of protein utilizing the principle of protein-dye binding. *Anal. Biochem.* **72:** 248–254.

Guarente L. 1983. Yeast promoters and *lacZ* fusions designed to study expression of cloned genes in yeast. *Methods Enzymol.* **101:** 181–191.

Miller J.H. 1972. *Experiments in molecular genetics.* Cold Spring Harbor Laboratory, Cold Spring Harbor, New York.

Rose M. and Botstein D. 1983. Construction and use of gene fusions to *lacZ* (β-galactosidase) that are expressed in yeast. *Methods Enzymol.* **101:** 167–180.

Plate Assay for Carboxypeptidase Y

PROCEDURE

This protocol was adapted from Jones (1991).

1. Grow colonies or patches on YPD plates (3 days for colonies and 1 day for patches are usually sufficient).
2. Carefully pour overlay solution over the surface of the plate to completely cover the cells.
3. After the agar hardens in 5–10 minutes, carefully flood the surface with *fresh* Fast Garnet GBC solution.
4. Allow color to develop for 5 minutes. Wild-type strains will turn red, and carboxypeptidase-Y-negative strains will appear yellow or pink.
5. Decant Fast Garnet solution to best observe developed color.

MATERIALS AND SOLUTIONS

Overlay solution for one plate

In glass or polypropylene tube, mix 2.5 ml of 1 mg/ml *N*-acetyl-DL-phenylalanine β-naphthyl ester in dimethylformamide with 4 ml of 0.6% molten agar. Hold at 50°C.

Fast Garnet GBC solution (use 5 ml/plate)

5 mg/ml Fast Garnet GBC (sulfate salt; Sigma F8761) in 0.1 M Tris-HCl (pH 7.4)

Dimethylformamide permeabilizes the cells. Carboxypeptidase Y in the cells cleaves the ester linkage in *N*-acetyl-DL-phenylalanine β-naphthyl ester. Free β-naphthol then reacts with the diazonium salt Fast Garnet GBC to produce an insoluble red dye.

REFERENCE

Jones E.W. 1991. Tackling the protease problem in *Saccharomyces cerevisiae. Methods Enzymol.* **194:** 428–453.

Random Spore Analysis

PROCEDURE

Sporulation

1. Patch out a single colony of the diploid to be sporulated onto YPD. It is best to spread the cells as thinly as possible. Incubate for 12–16 hours at 30°C. Growth for more than 16 hours will dramatically decrease the efficiency of sporulation.
2. In the morning, with a sterile dowel transfer a matchhead quantity of cells to a test tube containing 2.5 ml of sporulation medium and place on a rotor at 25°C.
3. Monitor the extent of sporulation by light microscopy. It can take from 2 to 10 days for more than 5% of cells to sporulate.

Random Spores

4. Transfer 1 ml of sporulated culture to a 15-ml conical polystyrene tube and collect the cells by centrifugation for 5 minutes in the clinical centrifuge.
5. Remove the supernatant completely and resuspend cells in 0.2 ml of sterile H_2O and 5 µl of β-glucuronidase (~500 units).
6. Incubate on a rotor for 1 hour at 30°C.
7. Add 0.1 ml (~0.15 g) of sterile 0.5-mm glass beads. Incubate on a rotor for 1 hour at 30°C.
8. Add 1 ml of sterile H_2O.
9. Vortex 1–2 minutes and check by light microscopy for complete disruption of asci.
10. Add 4 ml of sterile H_2O.
11. Make dilutions of 10^{-1}–10^{-3} in sterile H_2O. Plate 200 µl onto SC-arg with 60 µg/ml of canavanine.

MATERIALS AND SOLUTIONS

Sporulation medium
- 1% KOAc
- 0.025% glucose

β-glucuronidase (Sigma G7770)
Sterile 0.5-mm glass beads
SC-arg with 60 μg/ml of canavanine plates

Yeast Vital Stains

PROCEDURE

A. DAPI Stain of Nuclear and Mitochondrial DNA

Based on a method by Pringle et al. (1989).

SAFETY NOTES

DAPI is a possible carcinogen. It may be harmful if inhaled, swallowed, or absorbed through the skin. It may also cause irritation. Wear gloves, face mask, and safety glasses, and do not breathe the dust.

1. Pellet approximately 10^7 cells in a microcentrifuge tube (5-sec pulse) and resuspend in 70% ethanol.
2. Fix for 5 minutes or more and wash twice with H_2O.
3. Suspend cells in a small volume of 50 ng/ml DAPI (4′,6-diamidino-2-phenylindole; Sigma D9542 or Accurate Chemical and Scientific Corp.) in mounting medium. A stock of 1 mg/ml in H_2O can be stored at –20°C.
4. Observe with UV filter set.

Cells fixed in formaldehyde can also be stained with DAPI in mounting medium. DAPI can also be used as a vital stain of cells in growth medium at a concentration of approximately 1 μg/ml, but the background staining of the cell bodies is higher than with fixed cells. Staining can often be improved by including Triton X-100 (Sigma) to 0.1% in the medium.

B. Visualizing Mitochondria with $DiOC_6$ or $DiIC_5(3)$

Based on a method by Pringle et al. (1989).

To approximately 10^7 cells in growth medium, add $DiOC_6$ (3,3′-dihexyloxacarbocyanine iodide; Sigma D3652 or Molecular Probes) to a final concentration of 100 ng/ml (dilute $1/10^4$ from 1 mg/ml stock in ethanol, which is stable for months in the dark at –20°C), incubate for 5–10 minutes, and observe with a fluorescein filter set.

Alternatively, stain with $DiIC_5(3)$ (1,1´-dipentyl-3,3,3´,3´-tetramethylindocarbocyanine iodide; Molecular Probes) to a final concentration of 100 ng/ml (diluted from a 2.5 mg/ml stock solution in 100% ethanol), incubate for 5–10 minutes, and observe with a rhodamine filter set.

Note: The concentration of $DiOC_6$ and $DiIC_5(3)$ may need to be optimized; for example, at approximately 1 µg/ml all membranes appear to be stained.

C. Visualizing Vacuoles and Endocytic Compartments with FM4-64

Based on methods developed by Vida and Emr (1995) and Rieder et al. (1996).

1. Grow 5 ml of yeast to 2 x 10^7.
2. Harvest 1 ml and resuspend in 200 µl of YPD.
3. Add 2 µl of 8 mM FM4-64 (Molecular Probes #T-3166; make a 8 mM stock by dissolving 1 mg in 200 µl of H_2O) and incubate for 30–60 minutes in the dark at 30°C.
4. Pellet cells and wash with 1 ml of YPD.
5. Resuspend in 1 ml of YPD and chase for 20–40 minutes at 30°C.
6. Examine using the rhodamine filter set.

FM4-64 is a lipophilic styryl dye that is taken up by yeast cells via the endocytic pathway. It selectively stains the membranes of acidic compartments and can be used in pulse-chase experiments or in endocytic mutants to visualize the endocytic pathway in yeast. In fact, it has been used by Tom Vida to screen for endocytic mutants.

D. Calcofluor Staining of Chitin and Bud Scars

Based on a method by Pringle et al. (1989).

1. To approximately 10^7 cells in growth medium, add Calcofluor (Fluorescent brightener 28; Sigma F3397) to a final concentration of 100 µg/ml (dilute one-tenth from 1 mg/ml stock in H_2O, which is stable for weeks in the dark at –20°C).
2. Incubate 5 minutes or more, wash twice with H_2O, and observe with a DAPI-compatible filter set.

MATERIALS AND SOLUTIONS

Mounting medium

Dissolve 100 mg of *p*-phenylenediamine in 10 ml of PBS, adjust the pH to above 8.0 with 0.5 M Na carbonate buffer (pH 9.0), and bring the volume to 100 ml with glycerol. Add DAPI to 50 ng/ml. Mix thoroughly and store at –20°C. When mount get old, it turns brown.

REFERENCES

Pringle J.R., Preston R.A., Adams, A.E.M., Stearns T., Drubin D.G., Haarer B.K., and Jones E.W. 1989. Fluorescence microscopy methods for yeast. *Methods Cell Biol.* **31:** 357–435.

Rieder S.E., Banta L.M., Kohrer K., McCaffery J.M., and Emr S.D. 1996. Multilamellar endosome-like compartment accumulates in the yeast vps28 vacoular protein sorting mutant. *Mol. Biol. Cell.* **7:** 985–999.

Vida T.A. and Emr S.D.1995. A new vital stain for visualizing vacuolar membrane dynamics and endocytosis in yeast. *J. Cell Biol.* **128:** 779–792.

Yeast Immunofluorescence

PROCEDURE

This method is based on a protocol described by Pringle et al. (1989).

SAFETY NOTES

Formaldehyde is highly toxic and volatile. It is also a possible carcinogen. It is readily absorbed through the skin and is irritating or destructive to the skin, eyes, mucous membranes, and upper respiratory tract. Avoid breathing the vapors. Wear appropriate gloves and safety glasses and always use in a chemical fume hood. Keep away from heat, sparks, and open flame.

DAPI is a possible carcinogen. It may be harmful by inhalation, ingestion, or skin absorption. It may also cause irritation. Avoid breathing the dust and vapors. Wear appropriate gloves and safety glasses and use in a chemical fume hood.

The phenylenediamine in the mounting solution is a toxin and a potential carcinogen. It may be harmful by inhalation, ingestion, or skin absorption. Wear appropriate gloves and safety glasses and use in a chemical fume hood.

Preparation of Cells

1. Grow cells at the appropriate temperature (or nonpermissive temperature if applicable) to 2×10^7 in 25 ml of YPD. Add 17 ml of 10% formaldehyde (EM grade, Polysciences #04018) to a final concentration of 4% and incubate on the shaker at the temperature at which the cells were grown for 10 minutes.
2. Spin down the cells at 2000 rpm for 3 minutes and suspend in 6 ml of 40 mM KPO_4 (pH 6.5)/500 μM $MgCl_2$+4 ml of 10% formaldehyde. Incubate for 1 hour at 30°C.
3. Wash the cells twice in the previous buffer (no formaldehyde) and once in the same buffer containing 1.2 M sorbitol. Be very gentle with the cells. Resuspend in 0.5 ml of sorbitol buffer. At this point, the cells can be stored overnight at 4°C. Note that some reversal of formaldehyde cross-linking does occur if stored overnight.
4. Treat the cells with 30 μl of 10 mg/ml Zymolyase (100T) for between 5 and 30 minutes at 37°C. Examine cells on a phase microscope. If cells are dark and misshapen,

you have gone way too far; if bright and refractile, they probably need to be incubated longer. If their morphology still looks good but they are dull gray, they are just right. I suggest this be done as a time course comparing the Zymolyase-treated cells to a slide of cells before Zymolyase addition because it is the single most variable part of the procedure and unfortunately, the most critical.

5. Wash the cells once in sorbitol buffer and suspend in 100–500 μl of the same. *Be very gentle;* vortex at slow speeds to suspend the cells. Place on ice.

Preparation of Slides

6. Coat the wells of Teflon-faced slides (Polysciences #18357) with 20 μl of 0.1% polylysine (>400,000 MW) in water for 10 minutes at room temperature. Immediately before use, spin this solution for 10 minutes in a microfuge to remove dust and particulates; avoid using the solution from the bottom of the tube. In general, do high-speed spins of all solutions that go on the wells right before they are used to remove particulates. Incubate the slides in a moist chamber.
7. Wash the wells five times with 20 μl of clean H_2O that has been cleared of all particulates by high-speed centrifugation (10 min at 13,000 rpm) and dry the wells. These can be prepared in advance if kept dust-free.
8. Spot 20 μl of cell suspension on the wells and incubate for 10 minutes at room temperature. I suggest that you do each sample in duplicate. Aspirate off most of the liquid, but not all (do not dry the wells), and plunge the slide into a coplin jar that contains cold methanol for 6 minutes. Plunge into another coplin jar that contains cold acetone for 30 seconds. It is important that the jars and contents are very cold, so surround the coplin jars with dry ice in an ice bucket for 1 or 2 hours before this step. Change your methanol and acetone frequently and do not use alcohol-soluble ink to mark your slides.
9. When you remove the slide from the acetone, immediately place against a slanted, flat, warm, clean surface so that the acetone evaporates without the creation of condensation. I like to use the top of a warm water bath or heating plate. It can help to wick away the excess acetone from the bottom edge with a laboratory tissue as the slide dries. This process should only take a few seconds. Wear gloves because finger grease will make a mess of things. Also, wash the talcum off the gloves.

Staining Cells

10. Block the cells with PBS (pH 7.4)/0.5% BSA/0.5% ovalbumin by adding 20 μl per well. Spin this solution for 10 minutes in a microfuge immediately before use, and avoid using solution from the very bottom of the tube. Sometimes the inclusion of 0.5% Tween 20 can help the specificity of the antibody and should be includ-

ed in this block solution. Be forewarned that the Tween reduces the hydrophobicity of the Teflon and sample mixing can occur. Incubate for 1 hour at room temperature in a moist chamber. The cells can be incubated overnight at this step.

11. Incubate the cells in block solution containing antibody at the appropriate dilution (this may need to be determined experimentally). A dilution series of 1:100–1:10,000 is a good place to start. Incubate for 1 hour or longer at room temperature, depending on the antibody. Sometimes an overnight incubation is helpful. This is also a good stopping point because it allows ample time for weak primary antibodies to bind the antigens.
12. Wash the cells on the wells four times for 10 minutes with 20 µl/well of the block solution.
13. Incubate the wells in 20 µl/well with secondary antibody conjugate diluted in block solution for 1 hour at room temperature. You may need to determine the best concentration for your secondary antibody as well. Most secondary antibodies work well at 1:1,000 but this may need to be determined experimentally.
14. Wash as before. Aspirate most of last wash off the cells but do not dry; mount immediately. Mount by slopping the mounting solution over the slide (no bubbles) and gradually laying the cover slip down by the long axis. Lay paper towels over the slide and squeeze out the excess mount. While holding the slide down with the fingers of one hand, clean the slide with a laboratory tissue. Finally, seal the edges of the slide with nail polish. Allow to dry and store at –20°C until you are ready to view the results. Before you place the slide on the scope, completely clean any residual mount from the slide; the mounting solution is damaging to microscope objectives.

MATERIALS AND SOLUTIONS

Zymolyase 100T

Dissolve at 10 mg/ml in the sorbitol-containing phosphate buffer described in the protocol for 1 hour on ice with frequent vortexing, spin for 10 minutes in a microcentrifuge, remove to a fresh tube, and flash freeze with liquid nitrogen. Store at –80°C. Do not repeatedly freeze and thaw; always flash freeze.

Polylysine

Use only very high-molecular-weight polylysine (>300,000 MW; Sigma P5899). Dissolve as a 1% stock solution in ddH_2O. Flash freeze on liquid nitrogen in aliquots and store at –80°C. Avoid repeated freezing and thawing; always flash freeze.

DAPI

Stock solution should be at 1 mg/ml in water and stored at –20°C.

Mounting solution

Dissolve 100 mg of *p*-phenylenediamine in 10 ml of PBS, adjust the pH to above 8.0 with 0.5 M sodium carbonate buffer (pH 9.0), and bring the volume to 100 ml with glycerol. Add DAPI to 50 ng/ml. Mix thoroughly and store at –80°C. When the mount gets old, it turns dark brown and becomes autofluorescent; at this time, it should be disposed of and fresh mount should be made.

Note: Most antibodies require the methanol+acetone fix/denaturation step for their epitopes to be efficiently recognized. However, some antibodies do not require this step, thereby allowing one to avoid artifacts that are sometimes associated with the methanol/acetone treatment. If one can avoid the methanol/acetone fixation step, 0.1% Triton X-100 should be included in the block as well as antibody solutions to help permeabilize the cells.

REFERENCE

Pringle J.R., Preston R.A., Adams A.E.M., Stearns T., Drubin D.G., Haarer B.K., and Jones E.W. 1989. Fluorescence microscopy methods for yeast. *Methods Cell Biol.* **31:** 357–435.

Actin Staining in Fixed Cells

Phalloidin binds specifically to F-actin, and fluorescent-tagged phalloidin stains the actin skeleton in cells in a manner that is very close to the staining pattern seen using anti-actin antibody.

PROCEDURE

SAFETY NOTES

Formaldehyde is highly toxic and volatile. It is also a possible carcinogen. It is readily absorbed through the skin and is irritating or destructive to the skin, eyes, mucous membranes, and upper respiratory tract. Avoid breathing the vapors. Wear appropriate gloves and safety glasses and always use in a chemical fume hood. Keep away from heat, sparks, and open flame.

1. Grow 25 ml of yeast to 2×10^7 in a 500-ml baffled flask.
2. Add 17 ml of EM (electron microscope)-grade, 10% formadehyde to the media to a final concentration of 4%. Fix in the medium for 10 minutes at the growth temperature.
3. Spin down cells at 2000–3000 rpm for 5 minutes.
4. Fix cells in 6 ml of PBS plus 4 ml of 10% EM-grade formadehyde for 1 hour at room temperature.
5. Wash cells twice with 10 ml of PBS and reconstitute in 500 µl of PBS.
6. Remove 180 µl of cells and add 20 µl of rhodamine-phalloidin or fluorescein-phalloidin (the stock solution is 6.6 µM in methanol). Incubate for 1 hour in the dark. Vortex approximately every 15 minutes.
7. Wash the cells five times in 1 ml of PBS.
8. Suspend the cells in approximately 200 µl of immunofluorescence mounting solution and store at –20°C. The stained cells can be kept for a few months.
9. Visualize the cells by placing 1 µl of cell suspension under a standard size coverslip and waiting for capillary action to draw the liquid out to the edge of the coverslip, thus flattening the cells. Be sure that no dust is on the slide or coverslip.

REAGENTS

PBS

8 g of NaCl (137 mM final concentration)
0.2 g of KCl (2.7 mM final concentration)
1.44 g of Na_2HPO_4 (10.1 mM final concentration)
0.24 g of KH_2PO_4 (1.77 mM final concentration)
Dissolve in 1 liter total volume, adjust pH to 7.2, and sterilize filter.

Formaldehyde: Polysciences (10% EM-grade, catalog #04018).
Rhodamine-phalloidin: Molecular Probes, Inc. (#R-415), dissolved in methanol according to the manufacturer's instructions to approximately 6.6 µM and stored at –20°C.
Mounting solution
Dissolve 100 mg of *p*-phenylenediamine in 10 ml of PBS, adjust the pH to above 8.0 with 0.5 M sodium carbonate buffer (pH 9.0), and bring the volume to 100 ml with glycerol. Add DAPI to 50 ng/ml. Mix thoroughly and store at –80°C. When the mount gets old, it turns dark brown and becomes autofluorescent. At this point, it should be disposed of and fresh mount should be made.

PCR-mediated Gene Disruption

A. One-step PCR Gene Disruption

PROCEDURE

This protocol uses two primers tailed with approximately 50 nucleotides homologous to the gene of interest that target insertion of the polymerase chain reaction (PCR) product to that locus. Typically, the primers are designed to replace the open reading frame of a yeast gene with a selectable marker. Each primer ends with a universal sequence (see below) that is designed to amplify various selectable markers from plasmid templates developed by Wach (1996), Longtine et al. (1998), and Goldstein and McCusker (1999). The region of homology is short and thus the efficiency of gene replacement is low. However, since the selectable markers have no sequence homology with the yeast genome, the selectable markers usually target to the locus of interest. Note that if one tries to use this procedure with a selectable marker derived from the yeast genome, the PCR product will frequently mis-target to the marker locus (gene conversion) or elsewhere in the genome. Gene knockouts should always be done in diploids followed by sporulation and tetrad dissection to ensure that there is only one insertion of the marker and that additional mutations have not been induced at other loci.

1. Reaction mix

   10 ng of plasmid template DNA
   5 μl of 10x *Taq* buffer
   5 μl of 2 mM dNTP mix
   5 μl of 5 of μM primer 1
   5 μl of 5 μM primer 2
   0.5 μl of 10 mg/ml BSA
   1 μl of *Taq* polymerase
   H_2O to 50 μl

   Note: If the NAT cassette is to be amplified, because of a high GC content, the reaction must be supplemented with DMSO to 10%.

2. PCR cycle profile

   94°C for 4 minutes

   94°C for 1 minute
   55°C for 1 minute
   72°C for 1 minute per kilobase pair
   ——→ for 25 cycles

   72°C for 20 minutes

3. To clean the PCR, extract it with 50 μl of 1:1 phenol:chloroform, spin for 5 minutes at 13,000 rpm in a microcentrifuge, and remove the supernatant to a fresh tube. Add 5 μl of 3 M sodium acetate (pH 7.0) and 150 μl of 100% ethanol. Precipitate for approximately 1 hour at –20°C. Pellet the DNA at 13,000 rpm for 10 minutes, pour off the ethanol, and dry the pellet. Reconsitute the DNA in 25 μl of TE (pH 8.0) and transform 10 μl into yeast by a high-efficiency protocol (see Techniques and Protocols #1, High-efficiency Transformation of Yeast).

MATERIALS

Universal primer sequences

F1 5′-(gene-specific sequence)-cgg atc ccc ggg tta att aa-3′

R1 5′-(gene-specific sequence)-gaa ttc gag ctc gtt taa ac-3′

Plasmid templates

pFA6a-kanMX6 PCR will amplify the *kan*^{r} gene from *Escherichia coli* that confers resistance to G418, driven by the *Aphis gossypii TEF* promoter and terminator. See Wach (1996) and Longtine et al. (1998).

pFA6a-His3MX6 PCR will amplify the *his5*$^+$ gene from *Saccharomyces pombe* (which will complement a *S. cerevisiae his3Δ* allele) driven by the *A. gossypii TEF* promoter and terminator. See Longtine et al. (1998).

pAG25 PCR will amplify the *nat*^{r} gene from *Streptomyces noursei* that confers resistance to nourseothricin, driven by the *A. gossypii TEF* promoter and terminator. See Goldstein and McCusker (1999).

YPD+G418

Use a standard YPD recipe supplemented with a 200 mg/liter effective concentration of G418 disulfate salt (Sigma #G5013). The G418 powder can be added directly to the plate medium before pouring. Note that because of phenotypic lag, you must allow the cells to recover from the transformation for 4 hours– overnight before plating.

YPD+NAT

Dissolve the nourseothricin (Werner Bio Agents #5.1000) in ddH_2O to 100 mg/ml, sterilize filter, and store at –80°C. Add 1 ml of stock solution to a standard 1-liter

YPD recipe immediately before pouring the plates. Note that because of phenotypic lag, you must allow the cells to recover from the transformation for 4 hours–overnight before plating.

B. Gene Disruption by Double-fusion PCR

Protocol developed by Amberg et al. (1995).

Gene replacement in *S. cerevisiae* requires homologous recombination, which is greatly facilitated by long regions of homologous sequence. Frequently, short-tailed primers do not provide sufficient regions of homology for efficient and accurate gene replacement. In addition, if one wants to replace a gene with a marker that has homology elsewhere in the yeast genome, the procedure can be complicated by mistargeting the cassette to the marker locus. These problems can be greatly reduced if not eliminated by increasing the length of homologous regions at the ends of the disruption cassette. This can be achieved by using the double-fusion PCR procedure described below. Note that this same procedure can be adapted to the kanMX cassette by using the F1 and R1 primers discussed above and beginning primers 2 and 3 (see Fig. 1) with complementary sequences to the F1 and R1 primers.

PROCEDURE

1. In separate PCR reactions, amplify the upstream and downstream ends of the gene of interest with primers at least 200 bases apart. Primer 2 should begin with 24 nucleotides complementary to the m13 forward primer (5′-GTC GTG ACT GGG AAA ACC CTG GCG-3′) and primer 3 should begin with 24 nucleotides complementary to the m13 reverse primer (TCC TGT GTG AAA TTG TTA TCC GCT). PCR amplify the marker using the m13 forward and reverse primers. The Prakash and

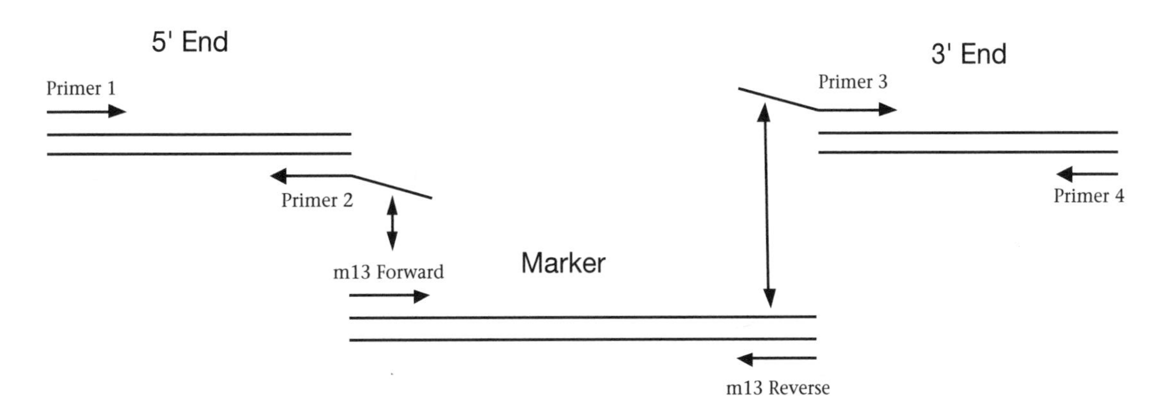

Figure 1. The primary PCRs required to construct a disruption cassette by double-fusion PCR.

Jones vectors citations are useful for the marker PCR. The conditions for these PCRs are as follows:

1–5 µl of yeast genomic DNA or 10 ng of plasmid template DNA
5 µl of 5 µM primer 1/m13 forward/primer 3
5 µl of 5 µm primer 2/m13 reverse/primer 4
5 µl of 10x *Taq* buffer
5 µl of 2 mM dNTPs
1 µl of *Taq*
H_2O to 50 ml

Leave for 4 minutes at 94°C, then do 25 cycles of 1 minute at 94°C, 1 minute at 55°C, and 3 minutes at 72°C. Finish with 20 minutes at 72°C and a 4°C soak.

2. Gel purify all of the PCR fragments on low-melt agarose.
3. Set up the first fusion PCR by melting the 5′ end fragment and marker fragment at 65°C, and adding to a PCR as follows:

   2.5 µl of each fragment
   5 µl of 5 µM primer 1
   5 µl of 5 µm m13 reverse primer
   5 µl of 10x *Taq* buffer
   5 µl of 2 mM dNTPs
   1 µl of *Taq*
   H_2O to 50 µl

   Leave for 4 minutes at 94°C, then do 25 cycles of 1 minute at 94°C, 1 minute at 55°C, and 3 minutes at 72°C. Finish with 20 minutes at 72°C and a 4°C soak.
4. Gel purify the product of the first fusion PCR on low-melt agarose.
5. Set up the second fusion PCR by melting the product of the first fusion PCR and the 3′ end fragments at 65°C, and adding to a PCR as follows:

   5 µl of each fragment
   10 µl of 5 µM primer 1
   10 µl of 5 µm primer 4
   10 µl of 10x *Taq* buffer
   10 µl of 2 mM dNTPs
   2 µl of *Taq*
   H_2O to 100 µl

   Leave for 4 minutes at 94°C, then do 25 cycles of 1 minute at 94°C, 1 minute at 55°C, and 3 minutes at 72°C. Finish with 20 minutes at 72°C and a 4°C soak.
6. Check for product by agarose gel electrophoresis. Extract the product with an equal volume of 1:1 phenol:chloroform, add 10 µl of 3 M sodium acetate (pH 7.0) and 300 µl of 100% ethanol, and precipitate the DNA for 1 hour or more at –20°C. Pellet the DNA at 13,000 rpm in a microcentrifuge for 10 minutes, pour off the supernatant, and dry the pellet. Reconstitute in 25 µl of TE (pH 8.0) and use 10 µl in a high-efficiency yeast transformation.

Note: Extracting the PCR products from the agarose will frequently increase the efficiency of the fusion PCRs. This is best done by purifying the PCR product on a gel through low-melting-temperature agarose, adding twice the volume of the gel slice TE (pH 8.0), extracting the excised DNA twice with phenol and once with chloroform, and following with ethanol precipitation.

MATERIALS AND SOLUTIONS

10x *Taq* buffer
- 0.2 M Tris (pH 8.3)
- 15 mM $MgCl_2$
- 0.25 M KCl
- 0.5% Tween 20

m13 forward primer
- 5´-cgc cag ggt ttt ccc agt cac gac-3´

m13 reverse primer
- 5´-agc gga taa caa ttt cac aca gga-3´

Gene-specific primer 2
- 5´-gtc gtg act ggg aaa acc ctg gcg(gene specific sequence)-3´

Gene-specific primer 3
- 5´-tcc tgt gtg aaa ttg tta tcc gct(gene-specific sequence)-3´

Plasmid templates
- pJJ215 Carries the *S. cerevisiae HIS3* gene cloned into the pUC18 polylinker (see Jones and Prakash 1990).
- pJJ242 Carries the *S. cerevisiae URA3* gene cloned into the pUC18 polylinker (see Jones and Prakash 1990).
- pJJ281 Carries the *S. cerevisiae TRP1* gene cloned into the pUC18 polylinker (see Jones and Prakash 1990).
- pJJ282 Carries the *S. cerevisiae LEU2* gene cloned into the pUC18 polylinker (see Jones and Prakash 1990).

Note: All of these markers are available in the opposite orientation (Jones and Prakash 1990).

REFERENCES

Amberg D.C., Botstein D., and Beasley E.M. 1995. Precise gene disruption in *Saccharomyces cerevisiae* by double fusion polymerase chain reaction. *Yeast* **11:** 1275–1280.

Goldstein A.L. and McCusker J.H. 1999. Three dominant drug resistance cassettes for gene disruption in *Saccharomyces cerevisiae. Yeast* **15:** 1541–1553.

Jones J.S. and Prakash L. 1990. Yeast *Saccharomyces cerevisiae* selectable markers in pUC18 polylinkers. *Yeast* **6:** 363–366.

Longtine M.S., McKenzie III A., Demarini D.J., Shah N.G., Wach A., Brachat A., Philippsen P., and Pringle J.R. 1998. Additional modules for versatile and economical PCR-based gene deletion and modification in *Saccharomyces cerevisiae. Yeast* **14:** 953–961.

Wach A. 1996. PCR-Synthesis of marker cassettes with long flanking homology regions for gene disruptions in *S. cerevisiae. Yeast* **12:** 259–265.

Yeast Colony PCR

PROCEDURE

1. Add 1 µl of Zymolyase 100T, 10 mg/ml to 9 µl of H_2O, and suspend a small yeast colony in this solution.
2. Add 40 µl of PCR mix to the suspended colony:

   5 µl of 5 mM gene-specific primer 1
   5 µl of 5 mM gene-specific primer 2
   5 µl of 10x *Taq* buffer
   5 µl of 2 mM dNTPs
   0.5 µl of 10 mg/ml BSA
   1 µl of *Taq*
   18.5 µl of H_2O

3. PCR cycle profile

   94°C for 4 minutes
   94°C for 1 minute
   55°C for 1 minute
   72°C 1 for minute per kb
   ——→ 35 cycles
   72°C for 10 minutes

4. Load 10 µl or more on a gel.

MATERIALS AND SOLUTIONS

10x *Taq* buffer
   0.2 M Tris (pH 8.3)
   15 mM $MgCl_2$
   0.25 M KCl
   0.5% Tween 20

Measuring Yeast Cell Density by Spectrophotometry

PROCEDURE

1. Take 1-ml samples of cells from the liquid medium in which they are growing, and place them in microfuge tubes. If necessary, dilute the cells to get an OD_{660} of less than 1.
2. Sonicate the cells to ensure an accurate reading of cell number, then transfer the samples to cuvettes. Prepare a blank using the growth medium in which the cells were grown. Use this sample to zero the absorbance before measuring the optical densities of the samples.
3. Measure the OD_{660} of each sample.
4. Calculate the cell density of haploid cells using Table 1 below (kindly provided by Lee Hartwell). Cell densities for diploids are half of those for haploids.

Note: There can be some variability among strains. For example, mutants that are abnormally large, such as some cell division cell (*cdc*) mutants, will scatter more light than wild-type cells at the same cell density. The table below refers to wild-type cells from strain A364A. There can also be some variability between spectrophotometers. It is advisable to carefully determine the cell density of your strain using a Coulter counter or a hemocytometer, and then measure the OD_{660} to construct a standard curve for your strain.

Table 1. Haploid yeast (A364a) cell density by OD_{660}

OD_{660}	# haploid cells x 10^7	OD_{660}	# haploid cells x 10^7	OD_{660}	# haploid cells x 10^7	OD_{660}	# haploid cells x 10^7
.000	.000	.500	.700	1.000	1.850	1.500	4.480
.010	.015	.510	.717	1.010	1.890	1.510	4.550
.020	.025	.520	.733	1.020	1.926	1.520	4.625
.030	.040	.530	.750	1.030	1.963	1.530	4.700
.040	.053	.540	.766	1.040	2.000	1.540	4.775
.050	.065	.550	.783	1.050	2.040	1.550	4.850
.060	.078	.560	.800	1.060	2.080	1.560	4.925
.070	.090	.570	.817	1.070	2.120	1.570	5.000
.080	.103	.580	.833	1.080	2.163	1.580	5.075
.090	.115	.590	.850	1.090	2.206	1.590	5.150
.100	.128	.600	.866	1.100	2.250	1.600	5.225
.110	.140	.610	.883	1.110	2.296	1.610	5.300
.120	.153	.620	.900	1.120	2.343	1.620	5.380
.130	.165	.630	.917	1.130	2.390	1.630	5.460
.140	.178	.640	.933	1.140	2.433	1.640	5.540
.150	.190	.650	.950	1.150	2.476	1.650	5.630
.160	.204	.660	.966	1.160	2.520	1.660	5.700
.170	.216	.670	.983	1.170	2.566	1.670	5.800
.180	.229	.680	1.000	1.180	2.613	1.680	5.890
.190	.241	.690	1.023	1.190	2.660	1.690	5.980
.200	.255	.700	1.046	1.200	2.706	1.700	6.070
.210	.268	.710	1.070	1.210	2.753		
.220	.280	.720	1.093	1.220	2.800		
.230	.293	.730	1.116	1.230	2.850		
.240	.305	.740	1.140	1.240	2.900		
.250	.319	.750	1.160	1.250	2.950		
.260	.330	.760	1.180	1.260	3.002		
.270	.342	.770	1.200	1.270	3.055		
.280	.356	.780	1.220	1.280	3.107		
.290	.370	.790	1.240	1.290	3.160		
.300	.385	.800	1.260	1.300	3.220		
.310	.399	.810	1.283	1.310	3.280		
.320	.412	.820	1.306	1.320	3.340		
.330	.426	.830	1.330	1.330	3.400		
.340	.440	.840	1.353	1.340	3.460		
.350	.455	.850	1.376	1.350	3.520		
.360	.470	.860	1.400	1.360	3.580		
.370	.484	.870	1.430	1.370	3.640		
.380	.499	.880	1.460	1.380	3.700		
.390	.514	.890	1.490	1.390	3.760		
.400	.530	.900	1.520	1.400	3.820		
.410	.547	.910	1.550	1.410	3.880		
.420	.564	.920	1.580	1.420	3.940		
.430	.580	.930	1.610	1.430	4.000		
.440	.600	.940	1.640	1.440	4.065		
.450	.617	.950	1.670	1.450	4.130		
.460	.633	.960	1.703	1.460	4.200		
.470	.650	.970	1.736	1.470	4.270		
.480	.666	.980	1.770	1.480	4.340		
.490	.683	.990	1.810	1.490	4.410		

Cell Synchrony

It is sometimes desirable to have an entire population of yeast cells that are growing synchronously or are arrested at a unique point in the cell cycle. One way to obtain synchrony is to use temperature-sensitive mutants that arrest at a specific point in the cell division cycle (*cdc* mutants). However, this requires constructing special strains for each experiment. An alternative approach is to use drugs or inhibitors to arrest the cells in the cell cycle.

PROCEDURE

α-factor

*MAT***a** cells will arrest at START in the cell cycle in response to the tridecapeptide mating pheromone α-factor that is commercially available (Sigma T6901). Store the peptide at 1 mg/ml in PBS at –20°C. *MAT***a** cells produce a protease (Bar1p) that destroys the α-factor; therefore, successful use of the α-factor in cell synchrony experiments requires that you accommodate for Bar1p activity. The degree of synchrony can be determined by the unique pear-shaped (schmoo) morphology that α-factor arrested cells adopt. Three approaches follow.

Dilute Concentration of Cells

This is useful when a small number of cells are needed (e.g., for microscopy).

1. Grow cells to 10^7 cells per ml in YPD.
2. Concentrate cells by centrifugation and twice wash with sterile water to remove Bar1p.
3. Resuspend cells at 10^4 cells per ml in YPD and add α-factor at 2 mg/ml.
4. Grow the cells for two generation times.
5. Concentrate cells by centrifugation and determine the percentage of cells that have arrested.

Low pH

The activity of Bar1p is pH dependent and can be partially suppressed at low pH.

1. Grow cells to 10^7 cells per ml in YPD.
2. Concentrate cells by centrifugation and twice wash with sterile water to remove Bar1p.
3. Resuspend cells at 10^6 cells per ml in YPD (pH 3.5) and add α-factor at 2 µg/ml.
4. Grow the cells for two generation times.
5. Determine the percentage of cells that have arrested.

bar1 Mutants

bar1 mutants lack the Bar1p protease and are very sensitive to pheromone, even at high cell densities.

1. Grow cells to 10^7 cells per ml in YPD.
2. Add α-factor to 50 ng/ml.
3. Grow the cells for two generation times.
4. Determine the percentage of cells that have arrested.

Reentering the Cell Cycle

To release cells from the α-factor-induced arrest, centrifuge the cells and wash twice with water. Resuspend the cells in medium containing 50 µg/ml of pronase (Sigma).

Hydroxyurea

Hydroxyurea is an inhibitor of the enzyme ribonucleotide reductase. Cells in the presence of the drug are unable to synthesize deoxyribonucleotides and therefore cannot complete DNA synthesis. Treating a culture with hydroxyurea results in a population of cells that arrest in the early stages of S phase. Cells arrested with hydroxyurea are large budded and have a single undivided nucleus and a short mitotic spindle. The effect of the drug is readily reversible.

1. Grow cells to 10^7 cells per ml in YPD.
2. Add hydroxyurea (Sigma), as a powder, to a final concentration of 0.2 M.
3. Grow the cells for two generation times.
4. Determine the percentage of cells that have arrested.

Reentering the Cell Cycle

To release cells from the hydroxyurea-induced arrest, centrifuge the cells and wash twice with water. Resuspend the cells in the appropriate medium.

Nocodazole

Nocodazole is an inhibitor of microtubule assembly, and cells in the presence of the drug cannot complete mitosis. Treating a culture with nocodazole results in a population of cells that arrest at M phase. Cells arrested with nocodazole are large budded and have a single undivided nucleus and no mitotic spindle. The nucleus is not located at the neck but is distributed randomly in the cell. The effect of the drug is readily reversible.

1. Grow cells to 10^7 cells per ml in YPD.
2. Add 10 µl of nocodazole (Sigma) per ml of cells from a stock solution of nocodazole that is 1.5 mg/ml in DMSO (final concentration 15 µg/ml nocodazole, 1% DMSO).
3. Grow the cells for two generation times.
4. Determine the percentage of cells that have arrested.

Reentering the Cell Cycle

To release cells from the nocodazole-induced arrest, centrifuge the cells and wash twice with water. Resuspend the cells in the appropriate medium.

Stationary Phase

Cells exit the cell cycle when they pass into stationary phase (G_0 arrest). When cells in stationary phase are diluted into fresh medium, they will reenter the cell cycle synchronously; however, the degree of synchrony is dependent on prior growth conditions and the strains. The utility of this method must be tested carefully before use.

1. Inoculate cultures at approximately 10^4 cells per ml in YP medium containing 2% raffinose and no glucose. Grow the cells for 48 hours at 30°C with constant aeration by shaking at 300 rpm in a rotary shaker. Be sure that the culture volume is 10% of the volume of the flask to assure optimal aeration.
2. Concentrate the cells by centrifugation and resuspend them at a concentration of 5×10^6 cells per ml in YPD medium. Be sure that the culture volume is 10% of the volume of the flask to assure optimal aeration. Assay cell synchrony by sampling every 15 minutes and determining the percentage of cells with small buds. Approximately 80% of the cells should initiate budding within a 15-minute interval after a 60- to 90-minute lag.

Chromatin Immunoprecipitation

PROCEDURE

1. Grow the yeast cells to be assayed. For each sample, grow 50–100 ml of cells to an OD_{600} of approximately 1.0 in a small culture flask.
2. Treat the cells with formaldehyde to cross-link proteins and DNA. Add formaldehyde to the cells to a final concentration of 1% (37% formaldehyde at 1:36) and maintain the cells for 10–120 minutes at room temperature, inverting occasionally.

   Note: The optimal fixation time varies depending on the protein of interest. This variable may need to be optimized for each specific assay.
3. Prepare lysis/immunoprecipitation (IP) buffer with protease inhibitors for the cell lysis step. I use 5–8 ml (more or less if desired; see steps 6 and 7) of lysis/IP buffer to collect each CHIP sample. The protease inhibitor stock solutions are 50x, so add 20 µl of each protease inhibitor per ml of lysis/IP buffer. Store the buffer in a cold room.

   Note: It is best to add PMSF last, *just before* using the buffer, since it is unstable in aqueous solutions, with a half-life of approximately 35 minutes at pH 8. The stocks are 50x, so add 20 µl of each protease inhibitor for each ml of buffer.
4. Quench the cross-linking reaction. Add 2.5 M glycine to a final concentration of 125 mM (a 20x dilution), and incubate the samples for 5 minutes at room temperature to quench the cross-linking reaction.
5. Wash the cells three times with TBS. Transfer the samples to GSA bottles or 50-ml polypropylene tubes to collect and wash the cells. Wash three times with ice cold TBS, using approximately 20 ml of TBS twice, and less in the final wash. The final wash should be performed in 50-ml tubes, removing the TBS as thoroughly as possible in preparation for the lysis step.
6. Lyse the cells with glass beads.
   a. Resuspend the cells in lysis/IP buffer. Resuspend the cells in 250 µl of ice cold lysis/IP buffer in the 50-ml tube.
   b. Add glass beads (0.5 mm) and vortex the cells. Add 3–6 ml of glass beads (for 1×10^9 to 2×10^9 cells) to the cells in lysis/IP buffer and use the S/P vortex mixer (it is the strongest machine) on the highest setting. After initially mixing the lysate and beads together, add enough beads so that there is approximately 1 ml of dry beads above the bead-lysate mixture in the tube. This may take an additional 1–2 ml of beads. The grinding action of the beads works

best if there is relatively little liquid in relation to beads. Vortex each sample 6–8 times for 30 seconds each time or 4–5 times for 1 minute each time. It is best to do this step entirely in the cold room. If doing it at room temperature, keep the cells cold by placing them on ice periodically.

Note: Examine the cells under the microscope to check for effective cell lysis. If the lysis is incomplete, repeat the vortexing step as needed.

7. Collect the crude cell lysate. Add 2–3 ml of fresh lysis/IP buffer to each sample, and pipette off the crude cell lysate from the beads. Collect the crude lysate into a 15-ml snap-cap tube. To collect the lysate, use a 1-ml automatic pipettor, and place the pipette tip at the bottom of the tube before releasing. Perform this step three times (using the desired total volume of lysis buffer) to collect as much lysate as possible. The volume collected will be slightly less than the total added, since some stays with the beads. Do not worry about transferring a few glass beads, because these will be removed in subsequent steps.

   Note: Take care not to cross-contaminate the lysate samples while pipetting or sonicating, since the final PCR analysis is very sensitive.

8. Sonicate the crude lysate to shear chromatin. Shear the sample chromatin by sonicating the suspension three times for 12–15 seconds each time. We use a Branson 250 sonifier with a microtip at power setting 3, 100% duty cycle. Between pulses, incubate the suspension on ice for at least 2 minutes. The average length of DNA postsonication should be 500 bp, with a range of 100–1000 bp.

   Note: Check the DNA fragment sizes after sonication with your sonifier to be certain that you have small DNA fragments. To reduce the chance of cross-contamination of samples, carefully clean the tip after each use of the sonifier.

9. Remove cell debris from the lysate by centrifugation. Spin the 15-ml tube at approximately 5000 rpm for a few minutes in the cold room tabletop centrifuge to remove the bulk of the cell debris. Transfer the supernatant into a 15-ml Corex tube for further centrifugation steps. Pellet the cell debris at 10.75000 in the Sorvall SS-34 rotor for 5 minutes at 4°C and decant the supernatant to a fresh tube. Repeat the centrifugation for 15 minutes at 4°C and again decant the supernatant to a fresh tube. The lysates should have a slightly milky appearance.

   Note: One can stop at this point by freezing the lysate samples at –80°C.

10. Normalize the amounts of protein in the lysate samples. Perform a total protein quantitation assay for each lysate sample. Since the lysates are very concentrated, measure total protein in a 10x diluted aliquot. Adjust the concentration of the lysates by adding lysis/IP buffer as needed, and use equivalent amounts of protein in each IP.

11. Remove total chromatin samples. Remove a 50-µl sample of each cell lysate and add 200 µl of TE/1% SDS to it. These are the total chromatin samples. The samples should contain sheared genomic DNA and are used to determine the equal presence of all DNA fragments before immunoprecipitation. Do not immunoprecipitate these samples, but take them through the post-IP reversal of cross-linking and all subsequent steps to provide input DNA for the control PCR.

12. Immunoprecipitate the protein of choice (one- or two-step IP). Perform three-step IPs, using cell lysates with 30 mg of total protein in each IP.

Preclearing. Preincubate the lysates with serum to remove nonspecific antigens. Use preimmune serum if available and "normal" serum if it is unavailable for this step.

Primary antibody. Add an appropriate amount of antibody to the lysate, and incubate it for one to several hours at 4°C.

No-antibody IP controls. Perform a no-antibody IP corresponding to each primary antibody IP, and follow with an agarose bead IP identical to that for the other IP samples. This control will indicate the level of nonspecific immunoprecipitation as a result of the beads or other factors.

Secondary antibody. Add 40 µl of a 50% suspension of protein A (for rabbit or rat antibodies) or protein G-agarose beads (for mouse antibodies). Incubate the samples for 1 hour at 4°C. This is a high-affinity binding step, and extended time is not likely to improve the IP.

13. Prepare lysis buffers with protease inhibitors for the post-IP washes. The post-IP wash step will require 2 ml of lysis buffer and 1 ml of lysis buffer/500 mM NaCl per sample. The protease inhibitor stock solutions are 50x, so add 20 µl of each protease inhibitor for each ml of buffer.

14. Wash the IP beads.

a. Remove the IP supernatant. Pellet the beads for 1 minute at 3000 rpm in an microcentrifuge (~1000*g*), and remove the supernatant. It may be useful to save this fraction for later analysis.

b. Wash the IP beads according to the following schedule. Between washes, spin the beads down for 1 minute at 1000*g* (as in step 13), and remove the wash liquid with a small pipette tip attached to an aspirator, taking care to avoid the pellet. The washes should be performed at room temperature with 1 ml of the indicated solution, for 3–5 minutes each time.

1. 2x with lysis buffer
2. 1x with lysis buffer/500 mM NaCl
3. 2x with IP wash solution
4. 1x with TE

15. Elute the immunoprecipitate with TES (TE/1% SDS).

a. Elute the immunoprecipitate from the antibody beads with 110 µl of TES (pH 8.0), incubating for 15 minutes at 65°C.

b. Pellet the beads for a few seconds at full speed (13,000 rpm) and transfer the eluate to a fresh tube. This is the first eluate fraction.

c. Wash the beads with 150 µl of TE/0.67% SDS, mix them, and pellet again. Remove the supernatant, which is a second eluate fraction, and add it to the first fraction.

   d. Spin the pooled eluate once more to eliminate any residual beads, and transfer it to a fresh tube. After centrifuging the combined eluate, remove 250 µl and avoid the ~10 µl left with the beads at the bottom of the tube.

16. Incubate all samples to reverse cross-linking. Incubate the eluates and the total chromatin samples for at least 6 hours at 68°C to reverse the cross-linking.

17. Treat the samples with proteinase K.

   a. Add TE to the samples. After the cross-linking has been reversed, add 250 µl of TE to get a total sample volume of 500 µl.

   b. Add glycogen and proteinase K to each sample. Add glycogen, a carrier for the DNA (add 5–10 µl of 10 mg/ml glycogen), and 100 µg of proteinase K (5 µl of 20 mg/ml stock).

   c. Incubate the samples for at least 2 hours at 37°C.

18. Add LiCl to the samples. Add 55 µl of 4 M LiCl to the solution. This salt will cause the DNA to precipitate at step 22.

19. Phenol/chloroform-extract the sample. Extract the samples with PCI (phenol:chloroform:isoamyl alcohol at 25:24:1). Extract the IP samples once with PCI followed by a chloroform:isoamyl alcohol extraction.

20. Precipitate the DNA.

   a. Add 1 ml of ethanol to each sample and mix. Store the samples in a cold place (I put them at -20°C or -70°C) to promote precipitation. Allow at least 15 minutes.

   b. Centrifuge the DNA at 0°C. Centrifugation at 12,000*g* for 10 minutes is usually sufficient to pellet DNA, but DNA of low concentration may require longer. Rinse the pellet with 75% ethanol, spin down and remove the rinse liquid, and allow the pellet to air-dry.

21. RNase treatment: Resuspend the pellet in 25–50 µl of TE containing 10 µg of RNase A. Incubate for 1 hour at 37°C.

22. Perform DNA amplification analysis of the sample DNA. Use specific primers to amplify the DNA target. Use primers from a nontarget locus as a negative control.

SAFETY NOTES

Chloroform is irritating to the skin, eyes, mucous membranes, and respiratory tract. It is a carcinogen and may damage the liver and kidneys. It is also volatile. Avoid breathing the vapors. Wear appropriate gloves and safety glasses and always use in a chemical fume hood.

Formaldehyde is highly toxic and volatile. It is also a possible carcinogen. It is readily absorbed through the skin and is irritating or destructive to the skin, eyes, mucous membranes, and upper respiratory tract. Avoid breathing the vapors. Wear appropriate gloves and safety glasses, and always use in a chemical fume hood. Keep away from heat, sparks, and open flame.

Isoamyl acetate is flammable and an irritant. It may be harmful by inhalation, ingestion, or skin absorption. Wear appropriate gloves and safety goggles. Do not breathe the vapors or mist. Keep away from heat, sparks, and open flame.

Phenol is extremely toxic, highly corrosive, and can cause severe burns. It may be harmful by inhalation, ingestion, or skin absorption. Wear appropriate gloves, goggles, and protective clothing, and always use in a chemical fume hood. Rinse any areas of skin that come in contact with phenol with a large volume of water and wash with soap and water; do not use ethanol.

MATERIALS AND REAGENTS

Yeast cells to be assayed (1 x 10^9 to 2 x 10^9 cells per sample)
Formaldehyde (37%)
2.5 M glycine in H_2O

TBS (Tris-buffered saline)	1 liter 10x stock:
20 mM Tris/HCl	200 ml of 1 M Tris/HCl (pH 7.6)
150 mM NaCl	300 ml of 5 M NaCl
	H_2O to reach 1 liter

Note: Dilute to working concentration and store in the cold room, because it is to be used cold.

Lysis/IP buffer	500 ml:
50 mM HEPES/KOH	25 ml of 1 M HEPES/KOH (pH 7.5)
140 mM NaCl	14 ml of 5 M NaCl
1 mM EDTA	1 ml of 500 mM EDTA
1% Triton X-100	50 ml of 10% Triton X-100
0.1% Na-deoxycholate	0.50 g of Na-deoxycholate
	H_2O to reach 500 ml

Protease inhibitors	Stock solutions:
1 mM PMSF	50 mM PMSF in isopropanol
1 mM benzamidine	50 mM benzamidine in H_2O
1 mg/ml of bacitracin	50 mg/ml of bacitracin in H_2O

Store these stock solutions in small aliquots (~0.5–1 ml) at –20°C. Each ml of lysis/IP buffer or lysis buffer/500 mM NaCl to be used will require 20 µl of each of these protease inhibitor stock solutions.

Lysis buffer/500 mM NaCl	250 ml:
50 mM HEPES/KOH	12.5 ml of 1 M HEPES/KOH (pH 7.5)
500 mM NaCl	25 ml of 5M NaCl
1 mM EDTA	0.5 ml of 500 mM EDTA (pH 7.5)
1% Triton X-100	25 ml of 10% Triton X-100
0.1% Na-deoxycholate	0.25 g of Na-deoxycholate

Note: Immediately before use, add protease inhibitors as you did for the lysis/IP buffer.

IP wash solution
- 10 mM Tris-HCl
- 0.25 M LiCl
- 0.5% NP-40
- 0.5% Na-deoxycholate
- 1 mM EDTA

250 ml:
- 2.5 ml of 1 M Tris-HCl (pH 8.0)
- 12.5 ml of 5 M LiCl
- 6.25 ml of 20% NP-40
- 1.25 g of Na-deoxycholate
- 0.5 ml of 500 mM EDTA

TES (TE/1% SDS)
- 50 mM Tris/HCl
- 10 mM EDTA
- 1% SDS

100 ml:
- 5 ml 1 M Tris/HCl (pH 8.0)
- 2 ml of 500 mM EDTA
- 5 ml of 20% SDS
- H_2O to reach 100 ml

TE/0.67% SDS
- 50 mM Tris-HCl
- 10 mM EDTA
- 0.67% SDS

100 ml:
- 5 ml of 1 M Tris-HCl (pH 8.0)
- 2 ml of 500 mM EDTA
- 3.35 ml of 20% SDS
- H_2O to reach 100 ml

TE (pH 8.0)
Branson sonifier 250
Primary antibody
Protein A or G-agarose beads
Protein A-sepharose bead buffer

- TE (pH 7.5)
- 0.1% BSA
- 0.1% sodium azide

50 ml:
- 47 ml of TE (pH 7.5)
- 2.5 ml of 2% BSA
- 0.5 ml of 10% sodium azide

Glycogen, 10 mg/ml in H_2O
Proteinase K, 20 mg/ml in H_2O
Store the stock solution in 110-µl aliquots at –20°C. 100 µl of proteinase K will treat 20 samples (see step 16).
4 M LiCl in H_2O
PCI (phenol:chloroform:isoamyl alcohol, 25:24:1)
Chloroform:isoamyl alcohol (24:1)
100% ethanol
75% ethanol/25% H_2O
RNase A
PCR reagents

Flow Cytometry of Yeast DNA

PROCEDURE

1. Grow and harvest the cells. Grow cells to 5 x 10^6 cells per ml. To collect the sample, spin down 10 ml of cells and wash them once in 5 ml of Tris buffer. Spin the cells down again and resuspend them in 1.5 ml of H_2O.
2. Fix the cells with ethanol. Add 3.5 ml of 100% ethanol to each 1.5-ml sample of cells to reach a final ethanol concentration of 70%. Leave the sample at room temperature for 1 hour. If necessary, store the sample at 4°C.
3. Wash out the ethanol and sonicate the cells. Spin the cells down and wash them in 5 ml of Tris buffer. Spin the cells down again and resuspend them in 5 ml of Tris buffer. Sonicate the sample to separate the cells (3x for 5–10 sec each time).
4. RNase-treat the cells. Dilute an RNase stock soution to 1 mg/ml (1x) in Tris buffer. Spin the cells down, remove the buffer, and resuspend them in 2 ml of 1x RNase solution. Incubate the sample for 1 hour at 37°C with shaking, then overnight at 4°C, if necessary. Complete RNase digestion is critical. This should be checked by looking at the cells with a fluorescence microscope (use rhodamine filters). The nuclear staining should be very bright and the cytoplasm should be devoid of stain except for mitochondrial DNA.
5. Treat the sample with pepsin. Spin down the cells and resuspend them in 1–2 ml of freshly prepared pepsin solution. Incubate the sample for 5 minutes at room temperature.
6. Stain the cells with SYTOX Green as described by Haase and Reed (2002). For each sample, prepare 0.5 ml of 50 mM Tris (pH 7.5) containing 2 μM SYTOX Green, which requires 1 μl of SYTOX Green per 2.5 ml of Tris. The final concentration of SYTOX in the sample is 1 μM.

REAGENTS AND MATERIALS

Tris buffer

50 mM Tris/HCl (pH 7.5)

10x RNase solution
  10 mg/ml of RNase
  100 mm of NaOAc

10 ml of 10x stock:
  100 mg of RNase
  333 µl of 3 M NaOAc (pH 5.0)
  9.7 ml of H_2O

Note: Boil the RNase stock solution for 30–60 minutes and store it at –20°C. To use the RNase, dilute it 10x in Tris buffer. Use RNase type 1-A, 5x crystallized.

Pepsin solution

10 ml:
  50 mg of pepsin
  9.45 ml of H_2O
  550 µl of 1 N HCl

Dissolve the pepsin in water before adding the HCl. Use only freshly prepared pepsin solution.

SYTOX Green
  5 mM in DMSO (Molecular Probes catalog # S-7020). Store in the dark at –20°C.

Staining buffer
  50 mM Tris/HCl (pH 7.5)

1 liter:
  Tris 6.0 g
  Adjust to pH 7.5.

REFERENCE

Haase S. and Reed S. 2002. Improved flow cytometric analysis of the budding yeast cell cycle. *Cell Cycle* **1:** 132–136.

Logarithmic Growth

It is often desirable to perform experiments on yeast cells that are in all possible stages of the cell cycle. It is necessary to have cells in the midlog stage of growth and not a culture with cells at stationary phase. Yeast cells grow with doubling times in the range of hours, and cells at stationary phase do not enter the cell cycle synchronously, except under carefully controlled growth conditions. Therefore, it is sometimes difficult and time-consuming to inoculate cells in the morning and be assured that they are normally distributed in the cell cycle. It is often necessary to inoculate cultures the day before to assure that they are at the right stage of growth for your use.

Logarithmic growth of a population of cells can be described mathematically as

$$N = N_0 e^{\ln 2\,(t/t_2)},$$

where

N = number of cells per ml at time = t
N_0 = number of cells per ml at $t = 0$
t = time in hours
t_2 = doubling time in hours

The cell density of wild-type yeast cells grown to stationary phase in YPD is approximately 2×10^8 cells/ml and the cell density of wild-type yeast cells grown to stationary phase in SC or HC medium is approximately 2×10^7 cells/ml.

Doubling times are approximately 2 hours at 23°C, 1.5 hours at 30°C, and 1 hour at 36°C.

Example:

It is 5:00 PM. You want to start an experiment at 9:00 AM tomorrow using cells at a density of 1×10^7 per ml at 30°C. You have a 5-ml culture of cells at saturation in YPD. How much do you have to dilute the culture to get the cells at the right density in the morning?

$N = 1 \times 10^7$
t = 16 hours
t_2 = 1.5 hours at 30°C

$1 \times 10^7 = N_0\ e^{\ln 2\ (16/1.5)}$
$1 \times 10^7 = N_0\ e^{(.69)(10.7)}$
$N_0 = 1 \times 10^7/1608$
$N_0 = 6218 = 6.2 \times 10^3$ cells per ml is the starting cell density that you want.

Dilute the cells: $2 \times 10^8/6218 = 3.2 \times 10^4$.

Therefore, 3.2×10^4 is the fold dilution of the saturated culture that is necessary to get 6.2×10^3 cells per ml.

For a 50-ml culture (50,000 µl), you need to inoculate with 1.6 µl of the overnight culture; i.e., 50,000/32,000 = 1.6.

EMS Mutagenesis

PROCEDURE

SAFETY NOTES

Ethylmethanesulfonate (EMS) is a volatile organic solvent that is a mutagen and carcinogen. It is harmful if inhaled, ingested, or absorbed through the skin. Discard supernatants and washes containing EMS in a beaker containing 50% sodium thiosulfate. Decontaminate all material that has come in contact with EMS by treatment in a large volume of 10% (w/v) sodium thiosulfate. Use extreme caution when handling. When using undiluted EMS, wear protective appropriate gloves and use in a chemical fume hood. Store EMS in the cold. DO NOT mouth-pipette EMS. Pipettes used with undiluted EMS should not be too warm; chill them in the refrigerator before use to minimize the volatility of EMS. All glassware coming in contact with EMS should be immersed in a large beaker of 1 N NaOH or laboratory bleach before recycling or disposal.

1. Grow an overnight culture to about 2 x 10^8 cells/ml.
2. Transfer two separate 1-ml samples of the same overnight culture to sterile microfuge tubes. Pellet the cells in a microfuge (10 sec at 5000*g*).
3. Discard the supernatant, resuspend the pellet in sterile distilled H_2O, and repellet. Repeat.
4. Resuspend the cells in 1 ml of sterile 0.1 M sodium phosphate buffer (pH 7).
5. Determine the exact cell density with a counting chamber.
6. Add 30 µl of ethylmethanesulfonate to one of the two tubes and disperse by vortexing vigorously. (The other tube will be the unmutagenized control.) Incubate both tubes for 1 hour at 30°C, with agitation.
7. Pellet the cells and remove the supernatant to a designated EMS waste container, then resuspend the cells in 200 µl of 5% sodium thiosulfate. Transfer to fresh tubes (discard the used tube in the EMS waste container).
8. Wash the cells twice with 200 µl of 5% sodium thiosulfate (each time discarding the supernatant in the EMS waste container). Resuspend the cells in 1 ml of sterile H_2O.
9. The cells may be plated directly, but in some cases it is important to provide a period of growth so that wild-type proteins in the cells can be replaced by the mutated versions, before exposing the cells to selective conditions. If outgrowth is performed, it is important to remember that this treatment results in the production of identical siblings in the culture.

Notes: This protocol will cause about 40–70% cell death in most haploid laboratory strains, but there is strain-dependent variation in sensitivity to the mutagen. This level of cell killing is commonly used in mutant hunts with haploid strains, but is not appropriate for all applications. Cells that survive this level of mutagenesis typically experience a 10^2–10^3 increase in the incidence of mutation relative to unmutagenized control cells (Lindegren et al. 1965). It is advisable to perform a pilot experiment with your strain to calibrate EMS-induced killing and mutation frequency. There are a number of simple tests that can be performed to assay mutation frequency. Perhaps the most useful is an assay for loss of function of the *CAN1* gene, which renders cells resistant to canavanine (Whelan et al. 1979). The generation of Can^r cells as a function of exposure to EMS can be monitored by plating mutagenized cells (along with unmutagenized control cells) on the canavanine-containing medium described in Appendix A. Another assay that monitors inactivation of a specific gene is the generation of cycloheximide-resistant cells (Kaufer et al. 1983). Cycloheximide resistance is nearly always the consequence of a specific transversion mutation in the *CYH2* gene. The assay is of limited utility for mutagens (such as EMS) that do not predominantly cause this type of mutation. Other assays that are commonly used to monitor mutagenesis include those in which inactivation of any of a few genes can yield the mutant phenotype. These include the generation of 5-FOA-resistant cells (loss of function of *URA3, URA5*, or other genes that may be involved in permeability; Boeke et al. 1986), growth on medium in which α-amino adipate is the sole nitrogen source (loss of *LYS2* or *LYS5* function; Chattoo et al. 1979), and the generation of red colonies (loss of *ADE1* or *ADE2* function; Jones and Fink 1982).

REFERENCES

Boeke J.D., LaCroute F., and Fink G.R. 1986. A positive selection for mutants lacking orotidine-5′-phosphate decarboxylase activity in yeast: 5′-fluoro-orotic acid resistance. *Mol. Gen. Genet.* **197:** 345–346.

Chattoo B.B., Sherman F., Azubalis D.A., Fjellstedt T.A., Mehnert D., and Ogur M. 1979. Selection of *lys2* mutants of the yeast *Saccharomyces cerevisiae* by the utilization of α-aminoadipate. *Genetics* **93:** 51–65.

Jones E.W. and Fink G.R. 1982. Regulation of amino acid and nucleotide biosynthesis in yeast. In *The molecular biology of the yeast* Saccharomyces: *Metabolism and gene expression* (ed. J.N. Strathern et al.). Cold Spring Harbor Laboratory, Cold Spring Harbor, New York.

Kaufer N.F., Fried H.M., Schwindinger W.F., Jasin M., and Warner J.R. 1983. Cycloheximide resistance in yeast: The gene and its protein. *Nucleic Acids Res.* **11:** 3123–3135.

Lindegren G., Hwang L.Y., Oshima Y., and Lindegren C. 1965. Genetical mutants induced by ethyl methanesulfonate in *Saccharomyces*. *Can. J. Genet. Cytol.* **7:** 491–499.

Whelan W.L., Gocke E., and Manney T.R. 1979. The *CAN1* locus of *Saccharomyces cerevisiae:* Fine-structure analysis and forward mutation rates. *Genetics* **91:** 35–51.

Tetrad Dissection

PROCEDURE

1. Sporulate cells on either plates or liquid medium. Examine the sporulated cultures to confirm that tetrads have been produced. Cultures showing less than 5% tetrads are difficult to dissect.
2. Prepare a fresh solution of Zymolyase T100 (ICN) (0.05 mg/ml in 1 M sorbitol) and place it at 4°C. Zymolyase contains β-glucuronidase, which cleaves bonds in the ascus coat, making it easier to break apart the ascospores. A less expensive alternative to Zymolyase is Gluculase (typically used as a 1:10 dilution of the stock), which generally yields less efficient digestion of the ascus coat, with greater spheroplasting of ascospores, than does Zymolyase.
3. If sporulation was performed in liquid (at a density of approximately 5 x 10^7 cells per ml), add 500 µl of the culture to a microfuge tube, centrifuge at 5000*g* for 10 seconds, and remove the supernatant. Gently resuspend the cell pellet in 50 µl of the Zymolyase solution and incubate at 30°C. If the culture was sporulated on plates, use the flat end of a sterile toothpick to transfer a dab of cells from the plate to a microfuge tube containing 50 µl of the Zymolyase solution. Suspend the cells by rotating the toothpick and incubate at 30°C.
3. Stop the Zymolyase digestion by placing the microfuge tube on ice and gently adding 150 µl of sterile H_2O. For most strains, an incubation time of approximately 10 minutes is appropriate, but some strains require significantly longer or shorter incubations. The appropriate incubation time can also vary for a specific strain as a function of sporulation time and conditions (e.g., liquid vs. solid sporulation medium). If you have not previously dissected your strain, a good strategy is to remove samples from the Zymolyase digestion at timed intervals (2–20 min). These samples can be examined microscopically to determine the ideal incubation time.
4. Very gently apply 10 µl of Zymolyase-treated cells as a streak across a YPD plate using a sterile loop. Alternatively, use a pipettor to transfer the cells to a tilted plate, allowing the droplet of cell suspension to run down the sloped agar surface leaving a stripe of Zymolyase-treated cells. With either method, care must be taken not to disrupt the tetrads that are now only tenuously intact because of the Zymolyase treatment. Generally, the stripe of Zymolyase-treated cells is made across the top or center of the plate. The cells must be accessible to the micromanipulator, and there must be sufficient space either above or below the stripe for placement of isolated ascospores.

5. Place the YPD plate containing the digested asci on the microscope stage. Be extremely careful not to break the microneedle. Adjust the stage so that a region of the plate without cells is in line with the objective lens. Focus the microscope on the surface of the plate and adjust the micromanipulator so that the needle appears in the center of the field. Clean the needle by dragging it through the agar.
6. Move the stage so that the cells are visible. Look for clusters of four ascospores that have a clear zone around them. Pick up the four spores with the microneedle and place them on the agar at least 5 mm from the stripe of Zymolyase-treated cells. Note the position on the mechanical stage. Pick up three spores and move the stage 5 mm away from the streak. Deposit the three spores and pick up two spores. Move the stage 5 mm away from the streak, deposit the two spores, and pick up one spore. Move the stage 5 mm and deposit the remaining spore. Move the microscope stage to the left or right 5 mm from the line of the four spores and select another four-spore cluster. Separate the spores as before by 5-mm intervals. By this method, ten tetrads can be dissected on each side of the YPD plate. Remove the YPD plate from the stage, taking care not to break the microneedle.
7. Incubate the plate for 2–3 days at 30°C.

DISSECTING TIPS

1. In most sporulated cultures there are many nonsporulated cells. Care must be taken to select only true tetrads and not groups of four cells that may resemble a tetrad. Spores are very spherical and slightly refractory. Nonsporulated cells may be less round and are often larger than spores. Select groups of four in which all four cells are in direct contact with one another. Avoid groups of four that fall apart extremely easily. The spores in true tetrads are usually mildly cohesive, even after Zymolyase treatment. Never attempt to guess which cells belong to the tetrad if you accidentally pick up more than four cells.
2. To separate spores that do not come apart easily in the dissection process, place the needle on the agar surface with its edge against the spores. Tap the side of the microscope gently with your finger. This should cause the needle to vibrate and often results in the separation of the spores.
3. If a spore is lost or becomes inextricably embedded in the agar, use a marking pen to indicate the position of that tetrad on the plate before moving on to the next tetrad. This can be done by rotating the objective lens to the side and making a mark on the plate corresponding to the position of the tetrad in question.
4. If the tetrad you would like to dissect is so crowded by other cells that you cannot pick it up without picking up other cells, create a clear zone around your tetrad by moving the surrounding cells with your needle. Once a clear zone has been created, remove all cells that might be attached to the needle by dragging it through the agar in a region of the plate that is free of cells. Then return to your tetrad and transfer it to a clear area for dissection.

5. Plates for dissection should be level, clear, thin, and dry. Pour the plates on a level surface so that if one side of the plate is in focus, the other will be as well. Certain brands of agar contain considerable debris, some of which resembles yeast spores. Bacto agar from Difco is one brand that is usually free of this type of debris. Since dissection microscopes usually focus on cells through the agar, thinner plates give clearer resolution of cells. Plates containing 25 ml of medium work well for most tetrad-dissection experiments.

   Cells are picked up on dissection needles in a small water droplet. For this reason, it is very difficult to pick up cells from a fresh, wet plate. Dry plates for 2 days at room temperature or overnight in a 37°C incubator before using them as dissection plates.

Making a Tetrad Dissection Needle

Microneedles attached to any of a variety of types of micromanipulators are useful in the dissection of tetrads, isolation of zygotes from populations of mating haploid cells, and manipulation of individual cells. Microneedles can be purchased from commercial sources (Cora Styles Needles 'N Blocks [www.tiac.net/users/cstyles/]). Alternatively, it is possible to make needles by gluing a thin glass filament with a flat end to a bent glass capillary pipette. Two methods for making your own needles are described below.

Method 1. Microneedles be made by drawing thin filaments from a 2-mm-diameter glass rod using a small gas flame (Scott and Snow 1978). The exact diameter is not critical, and investigators have different preferences. Spores are more readily picked up and transferred with microneedles having tips of larger diameters (100 μm), whereas manipulations in crowded areas having high densities of cells are more manageable with microneedles having smaller diameters (25 μm). A needle with a diameter of about 50 μm is an acceptable compromise. To draw out the needle, hold the glass rod over a low Bunsen burner flame until it is glowing orange. Simultaneously remove the rod from the flame while pulling the ends apart. With practice, it is possible to produce a glass thread using this method. Segments that appear to be the correct diameter are chopped into lengths of about 2 cm, placed on a glass slide that has been prewetted with water or saliva, then cut with a razor blade or glass coverslip to create microneedles 1 cM in length. The goal is to create microneedles with one perfectly smooth flat end as illustrated in Figure 1. The short segments are inspected with a microscope to determine whether they have the correct diameter and a flat end.

The microneedle mounting rod is made from a 2-mm-diameter glass. A 100-μl capillary pipette works well for this. Heat the glass rod over a low flame about 1 cM from its end, and when it becomes pliable, bend it to a right angle. A drop of Super Glue is applied to this end of the mounting rod, which is then touched to one of the segments of a glass filament as shown in Figure 1. The filament is carefully positioned so that it is at a right angle to the axis of the mounting rod and it has the correct length. The length of the perpendicular end should be compatible with the distance between the needle holder of the micromanipulator being used and the surface of the dissection plate. A too-short needle will not reach the surface of the medium, and one that is too long will dig into the surface of the medium. The microneedle holder is fitted into the micromanipulator after the glue has dried.

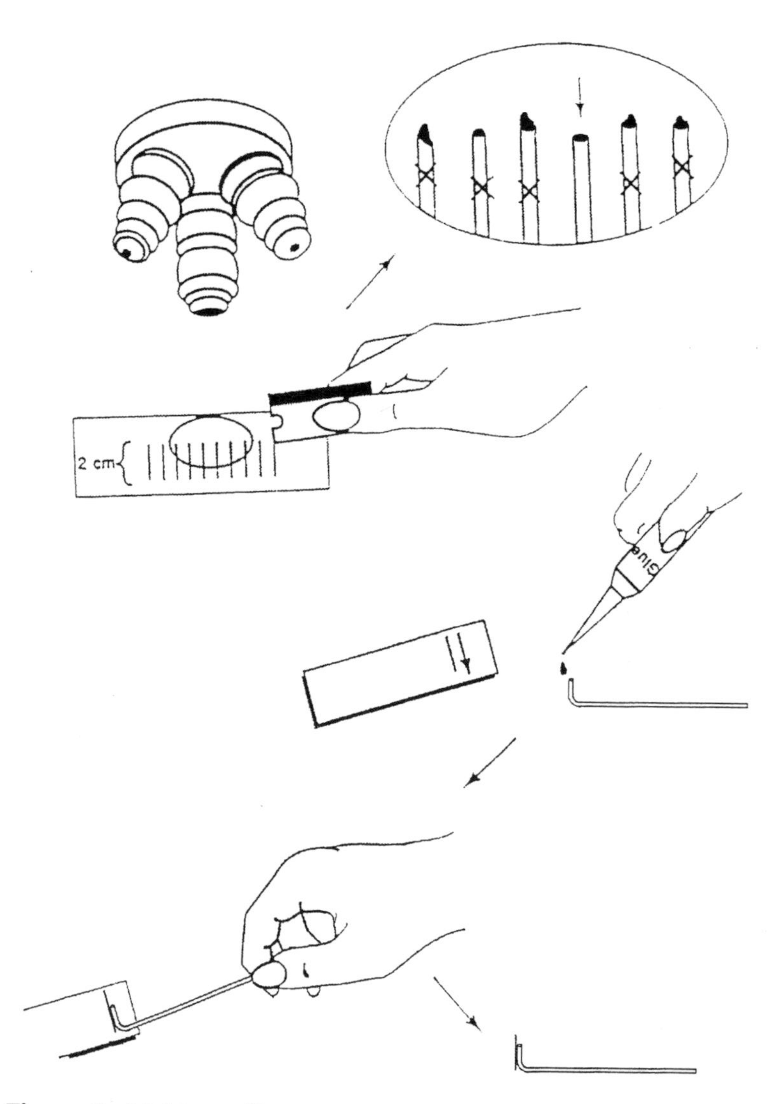

Figure 1. Making a dissection needle from drawn glass or optical fiber.

Method 2. This method is the same as Method 1 except that instead of pulling out a needle, commercially available fiber-optic glass (Edmond Scientific) of the correct diameter is used. The glass fiber is cut with a scissors to produce pieces that can be placed on a glass slide. These can then be cut with a razor blade or glass coverslip as described in Method 1 to create microneedles with flat ends. As described above, acceptable microneedles can then be mounted on a glass rod and attached to the micromanipulator.

REFERENCE

Scott K.E. and Snow R. 1978. A rapid method for making glass micromanipulator needles for use with microbial cells. *J. Gen. Appl. Microbiol.* **24:** 295–296.

Picking Zygotes

The unique morphology of zygotes makes it possible to identify them among populations of mating haploid cells. Zygotes can be easily separated away from nonmated cells using a micromanipulator. This method provides an alternative to the selection of diploid cells from a mating mixture by plating the mixture on a medium on which the diploid, but neither haploid parent, can grow.

Mating will be most efficient if the parent cells are from fresh cultures. Cells harvested from plates incubated for 1–2 days at 30°C generally work well. Cells from older cultures including those stored for several days at 4°C will mate, but the formation of zygotes will take longer than it would for fresh cultures.

PROCEDURE

1. Use sterile toothpicks to place equal amounts of the *MAT***a** and *MATα* parent cells (about the size of a 1- to 2-mm sphere) adjacent to one another on the surface of a YPD plate, then mix them well, creating a circle of about 5 mm in diameter.
2. Incubate the plate for 3 hours at 30°C.
3. Use a sterile toothpick to streak a sample of the mating mixture across the top of a fresh plate, such that individual cells are distinguishable when the plate is observed using a tetrad-dissecting microscope.
4. Zygotes form through the fusion of shmoos, the elongated haploid cells that have arrested in response to mating pheromones. These zygotes have a dumbbell shape that can be confused with large budded haploid cells in the mating mixture. Zygotes that have formed a medial bud have a distinctive three-lobed shape that allows them to be readily identified in the mating mixture.
5. Use methods described in Techniques and Protocols #22, Tetrad Dissection, to isolate these zygotes to a clean area of the plate using a microneedle designed for tetrad dissection.
6. Incubate the plate for 1–2 days at 30°C.
7. To confirm that the colonies arising from your isolated cells are indeed diploid, replica-plate them to media that will confirm the predicted phenotypes of your diploid strain. If the two haploid parents have no distinctive genetic markers, use mating test lawns (described in Experiment II) to confirm that your strain is a nonmater, the predicted mating type of a *MAT***a**/*MATα* diploid.

Determining Plating Efficiency

The plating efficiency of a strain measures the percentage of viable cells in a culture that are capable of forming colonies (sometime called colony forming units or c.f.u.). There are two simple procedures that can be used to determine plating efficiency.

PROCEDURE

Indirect Method

1. Determine the cell density in your culture using a Coulter counter by determining the optical density of the culture or by counting the cell number using a hemocytometer.
2. Determine the dilution factor necessary to dilute your cells to 10^3 cells per ml. Sonicate the cells to disperse clumps and dilute the cells in sterile H_2O.
3. Plate 0.2 ml of the diluted cells on a YPD plate, incubate for several days at the required temperature, and count the number of colonies. Determine the plating efficiency as the number of colonies divided by 200. Express the number as a decimal (100 colonies is a plating efficiency of 0.5).

Direct Method

1. Use a culture of cells at 10^6–10^7 cells per ml. Sonicate to disperse cells.
2. Spread 0.2 ml of the cells onto the surface of a YPD plate and allow the liquid to dry on the plate.
3. Incubate the plate for 16–24 hours at the desired temperature.
4. Observe the surface of the plate with a tetrad-dissection microscope. Viable cells will produce a small rounded microcolony of 50–100 cells. Inviable cells will not form a colony but will be a disorganized collection of 1–10 cells. Count the number of viable and inviable cells directly and determine the plating efficiency.

DNA Miniprep from E. *coli*

PROCEDURE

1. Inoculate a single colony into 3 ml of Terrific Broth containing 100 µg/ml of ampicillin. Incubate overnight at 37°C.
2. Pellet 1.5 ml of cells in a microfuge tube at 13,000 rpm for 1 minute and remove the supernatant by aspiration.
3. Add 100 µl of Soln #1 (50 mM glucose/10 mM EDTA/25 mM Tris [pH 8.0]). Eject the tip into the tube, and use the tip to loosen the pellet and get the cells off of the sides of the tube. Fully suspend the pellet by vortexing.
4. Add 200 µl of Soln #2 (0.2 N NaOH/1% SDS). Mix by inverting the tube four times. Incubate for 5 minutes on ice.
5. Add 150 µl of Soln #3 (3 M potassium/5 M acetate). Mix by shaking the tube up and down four times. Incubate for 5 minutes on ice.
6. Centrifuge for 5 minutes at 13,000 rpm.
7. Pour the supernatant into a microcentrifuge tube that contains 500 µl of phenol, vortex well, and centrifuge for 5 minutes at 13,000 rpm.
8. Remove 300 µl of the supernatant to a fresh tube, add 750 µl of 100% ethanol, vortex, and incubate for 2 minutes at room temperature.
9. Pellet the DNA at 13,000 rpm for 5 minutes. Pour off the supernatant and drain on a paper towel for approximately 5 minutes.
10. Reconstitute the DNA pellet in 50 µl of Soln #4 (TE [pH 8.0]+10 µg/ml RNase A).

MATERIALS AND SOLUTIONS

Terrific Broth

6 g of bacto-tryptone
12 g of yeast extract
2 ml of glycerol
450 ml of H_2O

Autoclave for 25 minutes, cool, and then add 50 ml of a sterile solution of 0.17 M KH_2PO_4/0.72 M K_2HPO_4.

Soln #1

228 µl of 40% glucose
200 µl of 0.5 M EDTA
250 µl of 1 M Tris (pH 8.0)
ddH_2O to 10 ml total volume
Store at 4°C.

Soln #2

400 µl of 5 M sodium hydroxide
1 ml of 10% SDS
8.6 ml of ddH_2O
Store at room temperature.

Soln #3

60 ml of 5 M potassium acetate
11.5 ml of glacial acetic acid
28.5 ml of ddH_2O
Store at 4°C.

Soln #4

1 ml of TE (pH 8.0) (10 mM Tris [pH 8.0]/1 mM EDTA)
1 µl of 10 mg/ml RNase A that has been boiled for 10 minutes

Preparing and Transforming Competent E. *coli*

PROCEDURE FOR THE "HANAHAN" PROTOCOL

Adapted by C. Davies and M. Tibbetts from Hanahan (1983).

Cell Preparation

1. Inoculate 1 ml of SOB with ten small fresh colonies of *E. coli* and vortex to mix.
2. Use this to inoculate 200 ml of SOB+12 ml of 5 M NaCl in a 2-liter, wide-bottom flask.
3. Grow overnight with moderate (275 rpm) shaking at 21–23°C (room temperature is fine). Note that the flask should be well ventilated (the doubling time is ~2–3 hours).
4. The next day, follow the density in the flask until the OD_{550} is 0.45–0.55.
5. Chill cells for 5–10 minutes on ice. Note that the cells can be stored on ice for more extended times.
6. Pellet cells at 3000–4000 rpm for 10 minutes at 4°C.
7. Gently resuspend the cell pellet in 20 ml of ice cold FCB.
8. Incubate for 15 minutes on ice.
9. Add 700 µl of freshly thawed dry DMSO. Add slowly while swirling the cell solution.
10. Incubate for 10 minutes on ice.
11. Add another 700 µl of DMSO as in step 9.
12. Incubate for 15 minutes on ice.
13. Aliquot cells into cold cryotubes (0.5- to 1.0-ml aliquots).
14. Flash freeze the aliquots on dry ice/ethanol or liquid nitrogen.
15. Store cells at –70°C.

Cell Transformation

1. Thaw an aliquot of cells slowly on ice.
2. Add DNA to 100 μl of cells.
3. Incubate for 20 minutes on ice.
4. Heat shock for 90 seconds at approximately 42°C.
5. Incubate for 10 minutes on ice.
6. Add 1 ml of LB and incubate for 30–60 minutes at 37°C.
7. (a) Plate aliquots or (b) spin at 4000 rpm for 4 minutes, remove all but approximately 150 μl of LB, gently resuspend the cells, and plate on LB+Amp or Tet.

MATERIALS AND SOLUTIONS

SOB

20 g of tryptone (2%)
5 g of yeast extract (0.5%)
0.58 g of NaCl (10 mM)
0.186 g of KCl (2.5 mM)
H_2O to 1 liter
Adjust pH to 7.5 with KOH (approximately 2 pellets)
Autoclave. Just before use, add
10 ml of 1 M $MgCl_2$ (10 mM)
20 ml of 1 M $MgSO_4$ (20 mM)

FCB

9.8 mg of potassium acetate (10 mM)
0.746 g of KCl (100 mM)
0.891 g $MnCl_2 \cdot 4H_2O$ (45 mM)
0.147 g of $CaCl_2 \cdot 2H_2O$ (10 mM)
80 mg of hexaminecobalt (III) chloride (3 mM)
10 ml of glycerol (10%)
H_2O to 100 ml
Filter sterilize; store at 4°C

Note: Purchase dry DMSO and place 1.5-ml aliquots into a –80°C freezer immediately upon arrival. Thaw one vial when ready to use.

THE "TSS" PROTOCOL

From Chung C.T. et al. (1989).

TSS 1x solution (may be made at 2x)

10% (w/v) PEG 3350 (PEG 8000 may be used but efficiency is slightly less)

5% DMSO (v/v)

20 mM Mg (use either $MgCl_2$ or $MgSO_4$; both are equally effective for transformation)

pH 6.5 (pH should be at or very close to 6.5 without any additions, check to confirm)

Procedure for Preparation of Chemically Competent Cells (~60x 100-μl aliquots)

This is the high-efficiency TSS protocol.

1. Select *E. coli* from a single colony.
2. Grow a 5-ml overnight culture at 37°C in LB.
3. Add 5 ml of culture to 600 ml and grow until OD_{595} is 0.3–0.4.
4. Chill for 15 minutes on ice.
5. Pellet *E. coli* by centrifugation at 4°C (e.g., 5 minutes in SLA-600 at 6000 rpm).
6. Pour off growth media and aspirate *all* media from cells.
7. Suspend cells in 6 ml of 1x TSS (ice cold).
8. Transfer 100-μl aliquots into cold 1.5-ml tubes (preincubated at –20°C is preferable).
9. Flash freeze in dry ice/MEOH bath or liquid nitrogen.
10. Store at –80°C.
11. Test and record transformation efficiency.

This procedure should result in *E. coli* (DH5α) with efficiencies of 1×10^7 to 5×10^7 transformants/μg DNA (plasmid).

Procedure for Preparation of "Lower"-efficiency Chemically Competent Cells (requires 2x TSS)

1. Follow guidelines above for culture of *E. coli.*
2. After *E. coli* reach the appropriate OD_{595}, cool for 15 minutes on ice.
3. Mix 1:1 with ice cold 2x TSS.
4. Flash freeze in dry ice/MEOH bath or liquid nitrogen.
5. Store at –80°C.
6. Test and record transformation efficiency.

Expect efficiencies of 1×10^6 to 5×10^6 transformants/μg DNA (plasmid). Efficiencies of TSS *E. coli* remain high with storage at –80°C, whereas $CaCl_2$ chemically competent cell transformation efficiencies decrease with continued storage at –80°C.

Transformation Protocol

1. Thaw competent cells on ice.
2. Add DNA (100 pg to 10 ng).
3. Mix and place on ice for 10 minutes.
4. Incubate for 10 minutes at room temperature.
5. Incubate for 10 minutes at 4°C (ice).
6. Add 1 ml of LB and incubate for 1 hour at 37°C.
7. Plate cells on appropriate selective media.

REFERENCES

Chung C.T., Niemela S.L., and Miller R.H. 1989. One-step preparation of competent *Escherichia coli:* Transformation and storage of bacterial cells in the same solution. *Proc. Natl. Acad. Sci.* **86:** 2172–2175.

Hanahan D. 1983. Studies on transformation of *Escherichia coli* with plasmids. *J. Mol. Biol.* **166:** 557–580.

Storing and Handling the Systematic Deletion Collection

Genetic and genomic analysis of budding yeast has developed dramatically in the past few years. There are publicly available collections of modified yeast strains for a variety of uses. Collections of deletion mutants are available in which almost every gene in the genome has been deleted by a one-step gene replacement with a *kanMX4* module. Conveniently, 20-bp unique oligonucleotide sequences have been inserted with the *kanMX4* module that serve as unique identifiers. These are called "bar codes." The unique bar codes can be amplified with a universal set of primers that amplify the 5′ bar code (uptag) and the 3′ bar code (downtag). The construction of the deletions is described in the *Saccharomyces* Deletion Consortium Web page (http://www-sequence.stanford.edu/group/yeast_deletion_project/deletions3.html). In other large collections of yeast strains, almost every open reading frame is tagged with glutathione *S*-transferase (GST), green fluorescent protein (GFP), or the tandem affinity purification (TAP) epitope. The strains are available individually or en masse from commercial vendors such as Open Biosystems or the American Type Culture Collection (ATCC).

The deletion collection is in an isogenic background and mutants are available that are isogenic with BY4741 (*MAT***a**), BY4742 (*MAT*α), and BY4743 (*MAT***a**/*MAT*α). Mutants that have deletions of nonessential genes are available as either haploids or heterozygous diploids. If you are buying an individual strain, purchase the diploid. Mutants with deletions of essential genes are only available as heterozygous diploids. If you are interested in a mutant with a deletion of a nonessential gene, you must dissect tetrads to obtain the mutant of the desired haploid genotype, isogenic with BY4741 or BY4742. Once the strain is obtained, it should be immediately frozen for further use. Deletion mutants are often compromised for growth, and prolonged passaging of haploid mutants can select for revertants, often aneuploids. The reduced fitness of the deletion mutant is often recessive and, therefore, the problem of changing genotype is minimized by using diploids. Recovering haploids after meiosis reduces the amount of times the haploids have been passaged. In addition, we recommend that you test for any other phenotype of your mutant to be sure that you have the right isolate. The G418 resistance is not sufficient, as all of the deletion mutants are thus marked. We suggest that you amplify the bar codes using the strategy described in Ooi et al. (2001). The resulting "uptag" and "downtag" DNA fragments are cloned and sequenced. The sequences of the bar codes can be obtained from the supplementary material of Ooi et al. (2001). This practice assures that you are working with the correct mutant.

Sometimes it is necessary to purchase an entire collection. For example, systematic genetic analysis (SGA) requires the entire deletion collection. The collection is sold in an ordered 96-well format. The collection shipped from the ATCC comes in 74 separate plates. The strains are shipped frozen in 150-μl glycerol stocks in sealed plates. The major concern in retrieving strains from the plates is cross contamination and great care must be taken to prevent this. Thaw the plates at room temperature and centrifuge them for 1 minute to remove any liquid that may be on the seal. Carefully remove the seal and transfer the strains from the original plate using sterile pins as described in Experiment #VI, Synthetic Lethal Mutants. Carefully reseal the plates with fresh seals (such as radiation-sterilized Nalgene Nunc 96-well seals, catalog #236366) and freeze the strains at –70°C. Use storage racks designed for microtiter plates. Create a database that indicates where the plates are stored. The strains can be pinned to a 384 format on omnitrays containing YPD+G418 by using the four guide holes in the colony copier as described in Experiment #VI. The first plate is pinned using hole A, the second using hole B, etc. After transferring to a 384 format, the plates are incubated for 2 days at 30°C. The arrayed colonies should be pinned to a 384-well microtiter dish containing 50 μl of YPD+200 μg/ml of G418. After 2 days growth at 30°C, add 25 μl of 45% glycerol, seal the plates, and freeze at –70°C using storage racks designed for microtiter plates. Create a database that indicates where the plates are stored. If possible, we advise that the collection be stored in duplicate in separate freezers. There are two internal checks for the integrity of the collection. The first is that the original 96-well microtiter plates have some wells that contain no yeast strains. They are an excellent indicator of the extent of cross contamination. If there is no growth in any of the wells that should be empty, cross contamination is minimal. The second test is the phenotype of large classes of mutants. Pin 384 plates to omintrays containing YPD, YPG, and SD agar. Petites will fail to grow on YPG and auxotrophs will fail to grow on SD. If all of the strains with the appropriate phenotypes are recovered, then cross contamination is minimal.

REFERENCE

Ooi S.L., Shoemaker D.D., and Boeke J.D. 2001. A DNA microarray-based genetic screen for nonhomologous end-joining mutants in *Saccharomyces cerevisiae*. *Science* **294:** 2552–2556.

Media

Media for Petri plates are prepared in 2-liter flasks, with each flask containing 1 liter of medium, which is sufficient for 30–40 plates. Unless otherwise stated, all components are autoclaved together for 15 minutes at 250°F (121°C) and 15 lb/square inch of pressure. Longer autoclaving of minimal media leads to hydrolysis of the agar, caramelized glucose, and mushy plates. If larger volumes are to be prepared, autoclave the salts, glucose, and agar separately for longer periods of time. The plates should be allowed to dry for 2–3 days at room temperature after pouring. The plates can be stored in sealed plastic bags for at least 3 months. The agar is omitted for liquid media. (For convenience, the final concentration of each component in the medium is listed in parentheses below.)

YPD (YEPD)

YPD is a complex medium for routine growth.

Bacto-yeast extract (1%)	10 g
Bacto-peptone (2%)	20 g
Glucose (2%)	20 g
Bacto-agar (2%)	20 g
Distilled H_2O	1000 ml

YPG (YEPG OR YEP-GLYCEROL)

YPG is a complex medium containing a nonfermentable carbon source (glycerol) that does not support the growth of ρ^- or *pet* mutants.

Bacto-yeast extract (1%)	10 g
Bacto-peptone (2%)	20 g
Glycerol (3% [v/v])	30 ml
Bacto-agar (2%)	20 g
Distilled H_2O	970 ml

YPAD (SLANT MEDIUM)

YPAD is a complex medium used for the preparation of slants. The adenine is added to inhibit the reversion of *ade1* and *ade2* mutants.

Bacto-yeast extract (1%)	10 g
Bacto-peptone (2%)	20 g
Glucose (2%)	20 g
Adenine sulfate (0.004%)	40 mg
Bacto-agar (2%)	20 g
Distilled H_2O	1000 ml

Dissolve the medium in a boiling-water bath. Dispense 1.5-ml portions with an automatic pipettor into 1-dram vials. Screw on the caps loosely and autoclave the vials. After autoclaving, incline the rack so that the agar is just below the neck of the vial. Tighten the caps after 1–2 days.

SYNTHETIC DEXTROSE (SD) MINIMAL MEDIUM

SD is a synthetic minimal medium containing salts, trace elements, vitamins, a nitrogen source (Bacto-yeast nitrogen base without amino acids), and glucose.

Bacto-yeast nitrogen base without amino acids (0.67%)	6.7 g
Glucose (2%)	20 g
Bacto-agar (2%)	20 g
Distilled H_2O	1000 ml

SUPPLEMENTED MINIMAL MEDIUM (SMM)

SMM is SD to which various growth supplements have been added. The specific constituents in SMM are defined in the Materials section of each experiment, where applicable. It is convenient to prepare sterile stock solutions by autoclaving for 15 minutes at 250°F (121°C). These solutions can then be stored for extensive periods. Some should be stored at room temperature to prevent precipitation, whereas the other solutions may be refrigerated. Wherever applicable, HCl salts of amino acids are preferred.

The medium should be prepared by adding the appropriate volumes of the stock solutions to the ingredients of SD medium and then adjusting the total volume to 1 liter with distilled H_2O. Threonine and aspartic acid solutions should be added separately to the medium after it is autoclaved.

Alternatively, it is often more convenient to prepare the medium by spreading a small quantity of the supplement(s) on the surface of an SD plate. The solution(s) should then be allowed to dry thoroughly onto the plate before inoculating with yeast strains.

Given below are the concentrations of the stock solutions, the volume of stock solution necessary for mixing 1 liter of medium, and the final concentration of each constituent in SMM. The volume of stock solution to spread on SD plates is also given.

Constituent	Stock concentration (g/100 ml)	Volume of stock for 1 liter of medium (ml)	Final concentration in medium (mg/liter)	Volume of stock to spread on plate (ml)
Adenine sulfate	0.2[a]	10	20	0.2
Uracil	0.2[a]	10	20	0.2
L-Tryptophan	1	2	20	0.1
L-Histidine HCl	1	2	20	0.1
L-Arginine HCl	1	2	20	0.1
L-Methionine	1	2	20	0.1
L-Tyrosine	0.2	15	30	0.2
L-Leucine	1	10	100	0.1
L-Isoleucine	1	3	30	0.1
L-Lysine HCl	1	3	30	0.1
L-Phenylalanine	1[a]	5	50	0.1
L-Glutamic acid	1[a]	10	100	0.2
L-Aspartic acid	1[a,b]	10	100	0.2
L-Valine	3	5	150	0.1
L-Threonine	4[a,b]	5	200	0.1
L-Serine	8	5	400	0.1

[a]Store at room temperature.
[b]Add after autoclaving the medium.

SYNTHETIC COMPLETE (SC) AND DROPOUT MEDIA

To test the growth requirements of strains, it is useful to have media in which each of the commonly encountered auxotrophies is supplemented except for the one of interest (dropout media). Dry growth supplements are stored premixed.

SC is a medium in which the dropout mix contains all possible supplements (i.e., nothing is "dropped out").

Bacto-yeast nitrogen base without amino acids (0.67%)	6.7 g
Glucose (2%)	20 g
Bacto-agar (2%)	20 g
Dropout mix (0.2%)	2 g
Distilled H_2O	1000 ml

Dropout mix:

Dropout mix is a combination of the following ingredients minus the appropriate supplement. It should be mixed very thoroughly by turning end-over-end for at least 15 minutes; adding a couple of clean marbles helps.

Adenine	0.5 g	Leucine	10.0 g
Alanine	2.0 g	Lysine	2.0 g
Arginine	2.0 g	Methionine	2.0 g
Asparagine	2.0 g	*para*-Aminobenzoic acid	2.0 g
Aspartic acid	2.0 g	Phenylalanine	2.0 g
Cysteine	2.0 g	Proline	2.0 g
Glutamine	2.0 g	Serine	2.0 g
Glutamic acid	2.0 g	Threonine	2.0 g
Glycine	2.0 g	Tryptophan	2.0 g
Histidine	2.0 g	Tyrosine	2.0 g
Inositol	2.0 g	Uracil	2.0 g
Isoleucine	2.0 g	Valine	2.0 g

HARTWELL'S COMPLETE (HC) MEDIUM

HC medium, used in Lee Hartwell's lab, is used in the same way as SC medium (see above); however, it uses different combinations of supplements for growth. This difference supports better growth of some strains. In addition, it is presented as a series of stock solutions, which gives very reproducible results between batches of media, and with a preference for easily making the six most common dropout media.

Stock Solutions

10x HC *(6 dropout amino acid liquid):*

Methionine	0.8 g
Tyrosine	2.4 g
Isoleucine	3.2 g
Phenylalanine	2.0 g
Glutamic acid	4.0 g
Threonine	8.0 g
Aspartic acid	4.0 g
Valine	6.0 g
Serine	16.0 g
Arginine	0.8 g

In a final volume of 4 liters; autoclave.

10x YNB:

Yeast nitrogen base (without amino acids and ammonium sulfate)	58 g
Ammonium sulfate	200 g

In a final volume of 4 liters; autoclave.

Amino acid solutions (note that these are NOT 10x solutions; see below):

Sterilize each solution by autoclaving. Keep the tryptophan in a light-sensitive bottle.

Uracil	1 g/liter
Adenine	1 g/liter
Lysine	10 g/liter
Tryptophan	10 g/liter
Leucine	20 g/liter
Histidine	10 g/liter

Recipe for Hartwell Complete Plates

Place 20 g of agar in 619 ml of distilled H_2O in a 2-liter flask. Autoclave for 20 minutes and then add the following:

20% glucose (sterile)	100 ml
10x YNB solution	100 ml
10x HC dropout 6 amino acid solution	100 ml
Uracil solution	35 ml
Adenine solution	20 ml
Lysine solution	12 ml
Tryptophan solution	8 ml
Leucine solution	4 ml
Histidine solution	2 ml

MAL INDICATOR MEDIUM

MAL indicator medium is a fermentation-indicator medium used to distinguish strains that ferment or do not ferment maltose. Because of the pH change, the maltose-fermenting strains will change the indicator yellow.

Bacto-yeast extract (1%)	10 g
Bacto-peptone (2%)	20 g
Maltose (2%)	20 g
Bromcresol purple solution (0.4% stock solution)	9 ml
Bacto-agar (2%)	20 g
Distilled H_2O	1000 ml

0.4% bromcresol purple solution:

Bromcresol purple	200 mg
100% ethanol	50 ml

GAL INDICATOR MEDIUM

GAL indicator medium is used for scoring the ability to ferment galactose.

Bacto-yeast extract (1%)	10 g
Peptone (2%)	20 g
Bacto-agar (2%)	20 g
Bromthymol blue solution (4 mg/ml stock solution)	20 ml
Distilled H_2O	880 ml

After autoclaving, add 100 ml of a filter-sterilized (0.2-µm filter) 20% galactose solution.

Bromthymol blue solution:

Bromthymol blue	400 mg
Distilled H_2O	100 ml

X-GAL INDICATOR PLATES FOR YEAST

5-Bromo-4-chloro-3-indolyl-D-galactoside (X-gal) does not work for yeast at the normal acidic pH of SD medium; therefore, a neutral pH medium is used. This is clearly a trade-off since many yeast strains will not grow well at this pH. For a first attempt at assessing β-galactosidase expression, this medium is worth a shot. The following are for 1 liter of X-gal indicator plates.

Solution I:

Mix

10x phosphate-buffer stock solution	100 ml
1000x mineral stock solution	1 ml
Dropout mix	2 g

Adjust the volume to 450 ml with distilled H_2O if the medium is to contain glucose, or to 400 ml if it is to contain galactose.

Solution II:

Mix in a 2-liter flask

Bacto-agar	20 g
Distilled H_2O	500 ml

Autoclave the solutions separately. After cooling to below 65°C, add the following to Solution I:

Glucose or other sugar to a final concentration of 2%	
X-gal (20 mg/ml dissolved in dimethylformamide)	2 ml
100x vitamin stock solution	10 ml
Any other heat-sensitive supplements	

Mix the solutions together and pour approximately 30 ml/plate.

10x phosphate-buffer stock solution:

KH_2PO_4 (1 M)	136.1 g
$(NH_4)_2SO_4$ (0.15 M)	19.8 g
KOH (0.75 N)	42.1 g
Distilled H_2O	1000 ml

Adjust the pH to 7 and autoclave.

SAFETY NOTE

Potassium hydroxide, KOH and KOH/methanol, is highly toxic and may be fatal if swallowed. It may be harmful by inhalation, ingestion, or skin absorption. Solutions are corrosive and can cause severe burns. It should be handled with great care. Wear appropriate gloves and safety goggles.

1000x mineral stock solution:

$FeCl_3$ (2 mM)	32 mg
$MgSO_4 \cdot 7H_2O$ (0.8 M)	19.72 g
Distilled H_2O	100 ml

Autoclave and store. This solution will form a fine yellow precipitate, which should be resuspended before use.

100x vitamin stock solution:

Thiamine (0.04 mg/ml)	4 mg
Biotin (2 µg/ml)	0.2 mg
Pyridoxine (0.04 mg/ml)	4 mg
Inositol (0.2 mg/ml)	20 mg
Pantothenic acid (0.04 mg/ml)	4 mg
Distilled H_2O	100 ml

Filter-sterilize using a 0.2-µm filter.

X-GAL PLATES FOR LYSED YEAST CELLS ON FILTERS

These plates are used for checking β-galactosidase activity in cells that have been lysed and are immobilized on 3MM filters.

Bacto-agar	20 g
1 M Na_2HPO_4	57.7 ml
1 M NaH_2PO_4	42.3 ml
$MgSO_4$	0.25 g
Distilled H_2O	900 ml

After autoclaving, add 6 ml of X-Gal solution (20 mg/ml in *N,N*-dimethylformamide).

SPORULATION MEDIUM

Strains will undergo several divisions on this medium and then sporulate after 3–5 days of incubation.

Potassium acetate (1%)	10 g
Bacto-yeast extract (0.1%)	1 g
Glucose (0.05%)	0.5 g
Bacto-agar (2%)	20 g
Distilled H_2O	1000 ml

Nutritional supplements are required for sporulation of auxotrophic diploids on sporulation medium. Supplements at the level of 25% of those used for SMM plates should be added when compounding the medium. Alternatively, supplements can be spread on the surface of the sporulation plates in the volumes listed for SMM. The liquid should be allowed to dry thoroughly onto the agar before inoculating it with yeast strains.

MINIMAL SPORULATION MEDIUM

*MAT***a**/*MAT*α diploid cells will sporulate on this medium after 18–24 hours without vegetative growth.

Potassium acetate (1%)	10 g
Bacto-agar (2%)	20 g
Distilled H_2O	1000 ml

Nutritional supplements are required for sporulation of auxotrophic diploids on sporulation medium. Supplements at the level of 25% of those used for SMM plates should be added when compounding the medium. Alternatively, supplements can be spread on the surface of the sporulation plates in the volumes listed for SMM. The liquid should be allowed to dry thoroughly onto the agar before inoculating it with yeast strains.

LOW-pH BLUE PLATES

These plates are used for testing killer phenotype.

Bacto-yeast extract (1%)	6 g
Bacto-peptone (2%)	12 g
Glucose (2%)	12 g
Bacto-agar (2%)	12 g
Distilled H_2O	533 ml

Autoclave the above ingredients and add the following solutions:

Methylene blue in sterile H_2O	5 ml
Phosphate-citrate buffer (sterile)	67 ml

Methylene blue in sterile H_2O:

Methylene blue	20 mg
Sterile H_2O	5 ml

Phosphate-citrate buffer for low-pH medium:

Citric acid	14.07 g
K_2HPO_4	18.96 g
Distilled H_2O	67 ml

Adjust the pH to 4.5 using solid K_2HPO_4 or citric acid. Sterilize by autoclaving.

DRUG SELECTION MEDIA

5-Fluoro-orotic Acid Medium

5-Fluoro-orotic acid (5-FOA) can be used to select for mutant cells that fail to utilize orotic acid as the source of the pyrimidine ring. Wild-type cells convert 5-FOA to 5-fluoro-orotidine monophosphate by conjugation to phosphoribosyl pyrophosphate (PRPP) and subsequently decarboxylate it to form 5-fluoro-uridine monophosphate (5-FUMP). These two steps are catalyzed by the products of the yeast genes *URA5* and *URA3*, respectively. Inevitably, fluorodeoxyuridine formed later is a potent inhibitor of thymidylate synthetase and thereby quite toxic to the cell. The two steps of de novo synthesis of uridine that are required to convert 5-FOA to 5-FUMP can be mutated to block utilization, as long as uracil is provided to allow formation of UMP via the salvage pathway. Therefore, both *ura3*⁻ and *ura5*⁻ mutants can grow on 5-FOA-containing medium (Boeke et al. 1984). In practice, only *ura3*⁻ mutants appear to be uracil auxotrophs. The enzyme that catalyzes conjugation of uracil to PRPP can utilize orotic acid as a substrate at some level, allowing *ura5*⁻ mutants to grow slowly in the absence of uracil.

Bacto-yeast nitrogen base (0.67%)	6.7 g
Dropout mix ura⁻ (0.2%)	2 g
Glucose (2%)	20 g
Uracil (50 μg/ml)	50 mg
5-FOA (0.1%)	1 g
Distilled H_2O	500 ml

Dissolve the above and filter-sterilize using a 0.2-μm filter.

Autoclave the agar separately:

Bacto-agar (2%)	20 g
Distilled H_2O	500 ml

Mix the two solutions after cooling the agar to approximately 80°C. Pour into Petri dishes (25 ml/dish).

5-FOA Medium a la HC

Make 1 liter of HC plates as described above, let flask cool to 50°C, then add 1.0 g of 5-FOA.

α-Aminoadipate Plates

Wild-type strains are unable to utilize high levels of α-aminoadipate (αAA) as their sole nitrogen source because it is converted into a toxic intermediate by the normal lysine anabolic pathway (Chattoo et al. 1979; Zaret and Sherman 1985). This medium is frequently used in the selection of mutations in the *LYS2* and *LYS5* genes.

Bacto-yeast nitrogen base without amino acids or ammonium sulfate (0.16%)	1.6 g
Glucose (2%)	20 g
Lysine (30 mg/liter)	30 mg
Bacto-agar (2%)	20 g
Distilled H_2O	960 ml

Autoclave and add 40 ml of a 5% αAA solution.

5% αAA:

α-Aminoadipic acid	2 g
Distilled H_2O	40 ml

Mix and adjust the pH to 6 with 10 N KOH to allow dissolution. Filter-sterilize using a 0.2-µm filter before adding to the autoclaved ingredients.

SAFETY NOTE

Cycloheximide may be fatal if inhaled, ingested, or absorbed through the skin. Wear appropriate gloves and safety glasses and use in a chemical fume hood.

Cycloheximide

Cycloheximide resistance can arise in a number of different genes, but resistance to high levels ordinarily occurs because of rare mutations at the *cyh2* locus, which encodes the L29 ribosomal subunit. Resistance to cycloheximide is recessive, presumably because the sensitive ribosomes remain bound to the mRNA and block all further elongation.

Cycloheximide can be used in either YPD or synthetic media. A final concentration of 10 mg/liter should be used for YPD and 3 mg/liter for SD, SC, and YPG. A stock solution is prepared by dissolving 100 mg of cycloheximide in 10 ml of distilled H_2O and then filter-sterilizing (0.2-µm filter). The stock solution can be stored at 4°C. Appropriate volumes can be added to media after autoclaving.

Canavanine

Canavanine is an analog of arginine. Both are imported into the cell via the same high-affinity permease, which is encoded by the *CAN1* locus. High-level resistance to canavanine occurs exclusively because of mutation at this locus, but low-level resistance can arise at a number of other loci.

Because canavanine is a competitive inhibitor, arginine must be excluded from media used for testing sensitivity to the drug. Canavanine resistance must also be scored under high-nitrogen conditions, such as those provided by SD or SC medium, since the *CAN1* permease will then provide the only entry route to the cell for arginine and canavanine. In the presence of low-nitrogen conditions—effectively those provided by YPD medium—the general amino acid permease (*GAP*) system is induced and arginine and canavanine can also be taken up by this route. In addition, Can^R Arg^- auxotrophs are viable on YPD but are inviable on synthetic media because they are unable to take up arginine.

Canavanine sulfate is typically made up as a filter-sterilized (0.2-μm filter) 20-mg/ml stock solution in distilled H_2O. It is stored at 4°C and added to SD or SC-arg medium after autoclaving. A concentration of 60 mg/liter is typically used for scoring and selecting canavanine resistance.

REFERENCES

Boeke J.D., LaCroute F., and Fink G.R. 1984. A positive selection for mutants lacking orotidine-5′-phosphate decarboxylase activity in yeast: 5-Fluoro-orotic acid resistance. *Mol. Gen. Genet.* **197:** 345–346.

Chattoo B.B., Sherman F., Azubalis D.A., Fjellstedt T.A., Mehnert D., and Ogur M. 1979. Selection of *lys2* mutants of the yeast *Saccharomyces cerevisiae* by the utilization of α-aminoadipate. *Genetics* **93:** 51–65.

Zaret K.S. and Sherman F. 1985. α-Aminoadipate as a primary nitrogen source for *Saccharomyces cerevisiae* mutants. *J. Bacteriol.* **162:** 579–583.

Appendix B

Stock Preservation

Yeast strains can be stored indefinitely in 25% (v/v) glycerol at a temperature of –60°C or less (Well and Stewart 1973). Yeast tends to die if stored at temperatures above –55°C. Many workers use 2-ml vials (35 x 12 mm) containing 1 ml of sterile 25% (v/v) glycerol. The strains are grown on the surfaces of YPD plates. The yeast is then scraped up with sterile applicator sticks or toothpicks and suspended in the glycerol solution. The caps are tightened and the vials shaken before freezing. The yeast can be revived by transferring a small portion of the frozen sample onto a YPD plate.

Yeast strains can be stored for up to 6 months at 4°C on slants prepared with YPAD medium. This method of storage is convenient since the slants take up little space, do not dry out, and contain excess adenine to prevent toxicity as a result of the red pigment produced by certain *ade*$^-$ mutants. Slants are also a useful means of sending strains to colleagues.

REFERENCE

Well A.M. and Stewart G.G. 1973. Storage of brewing yeasts by liquid nitrogen refrigeration. *Appl. Microbiol.* **26:** 577.

Appendix C

Yeast Genetic and Physical Maps

Figures I–XVI on the following pages (214–219) depict the genetic and physical maps, and their correlations, of the 16 *Saccharomyces cerevisiae* chromosomes. A parallel comparison of the physical map (left, in kilobase pairs) and the genetic map (right, in centimorgans) of each of the 16 chromosomes is illustrated. The information in these figures is available on the *Saccharomyces* Genome Database (SGD) http://www.pathway.yeastgenome.org/. The physical map consists of shaded boxes that indicate open reading frames (ORFs). ORFs on the Watson strand (left telomere is the 5′ end of this strand) are shown as light gray boxes, those on the Crick strand as dark gray boxes. Where it has been defined, the gene name of an ORF is indicated. The genetic map is based on data collected since 1991 by the SGD project, as well as earlier data. Horizontal tick marks on the right of the genetic map line indicate positions of genes. Lines connect genetically mapped genes with their ORF on the physical map. A single name is listed for known synonyms. (Reprinted, with permission, from Cherry et al. 1997 [©Macmillan Magazines Ltd.]).

REFERENCE

Cherry J.M., Ball C., Weng S., Juvik G., Schmidt R., Adler C., Dunn B., Dwight S., Riles L., Mortimer R.K., and Botstein D. 1997. Genetic and physical maps of *Saccharomyces cerevisiae*. *Nature* (suppl.) **387:** 67–73.

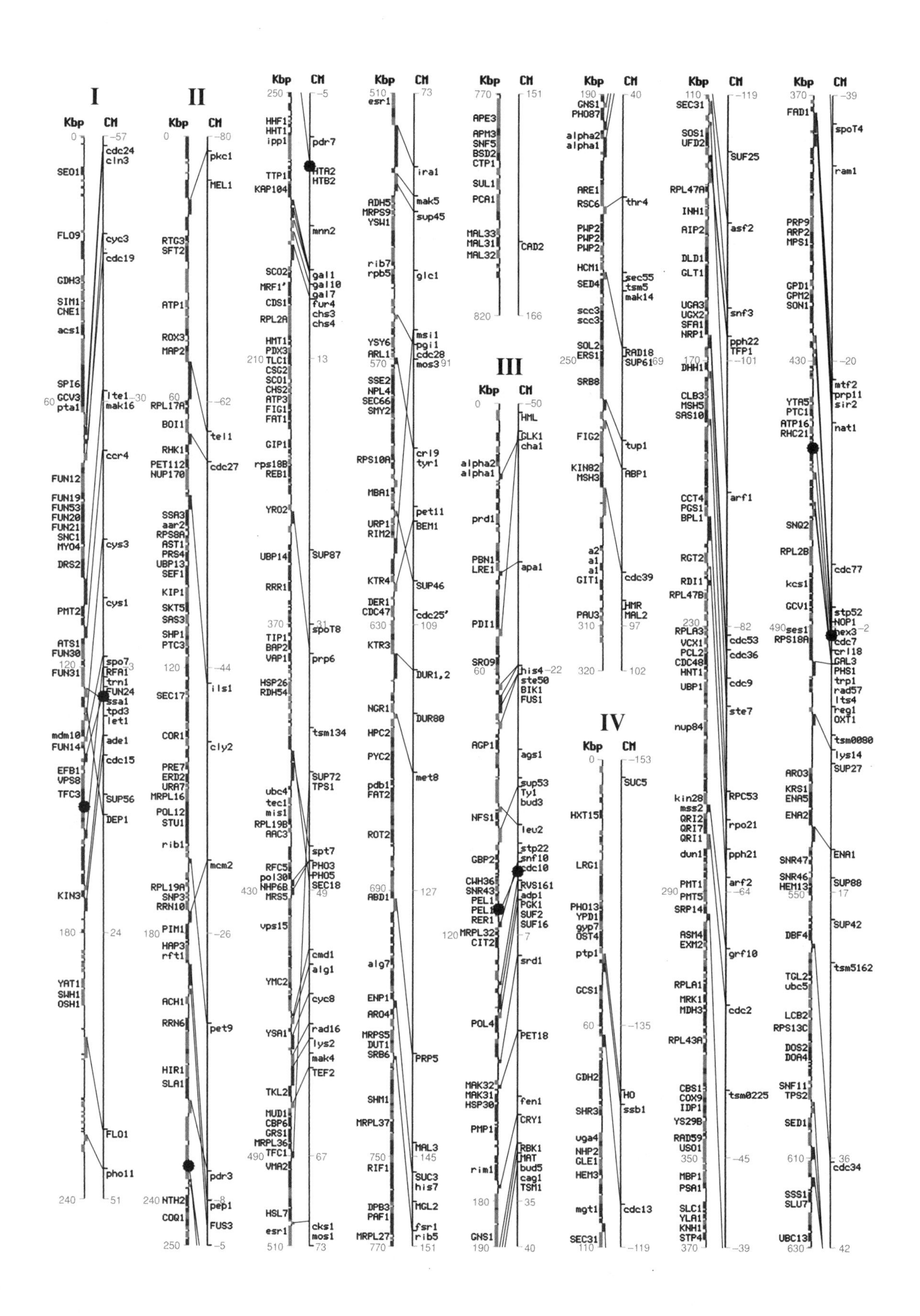
I
II
III
IV
Kbp
CM

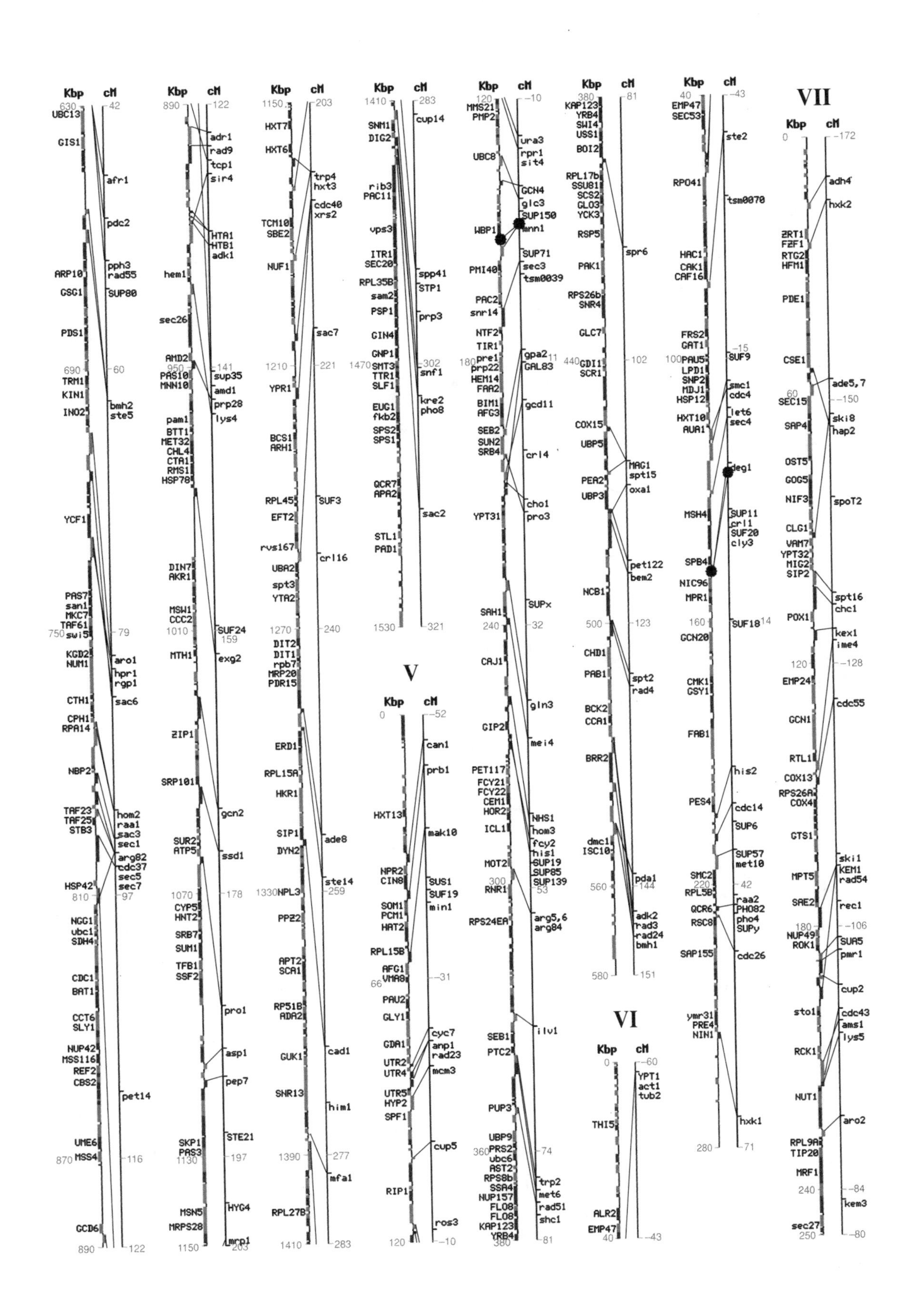

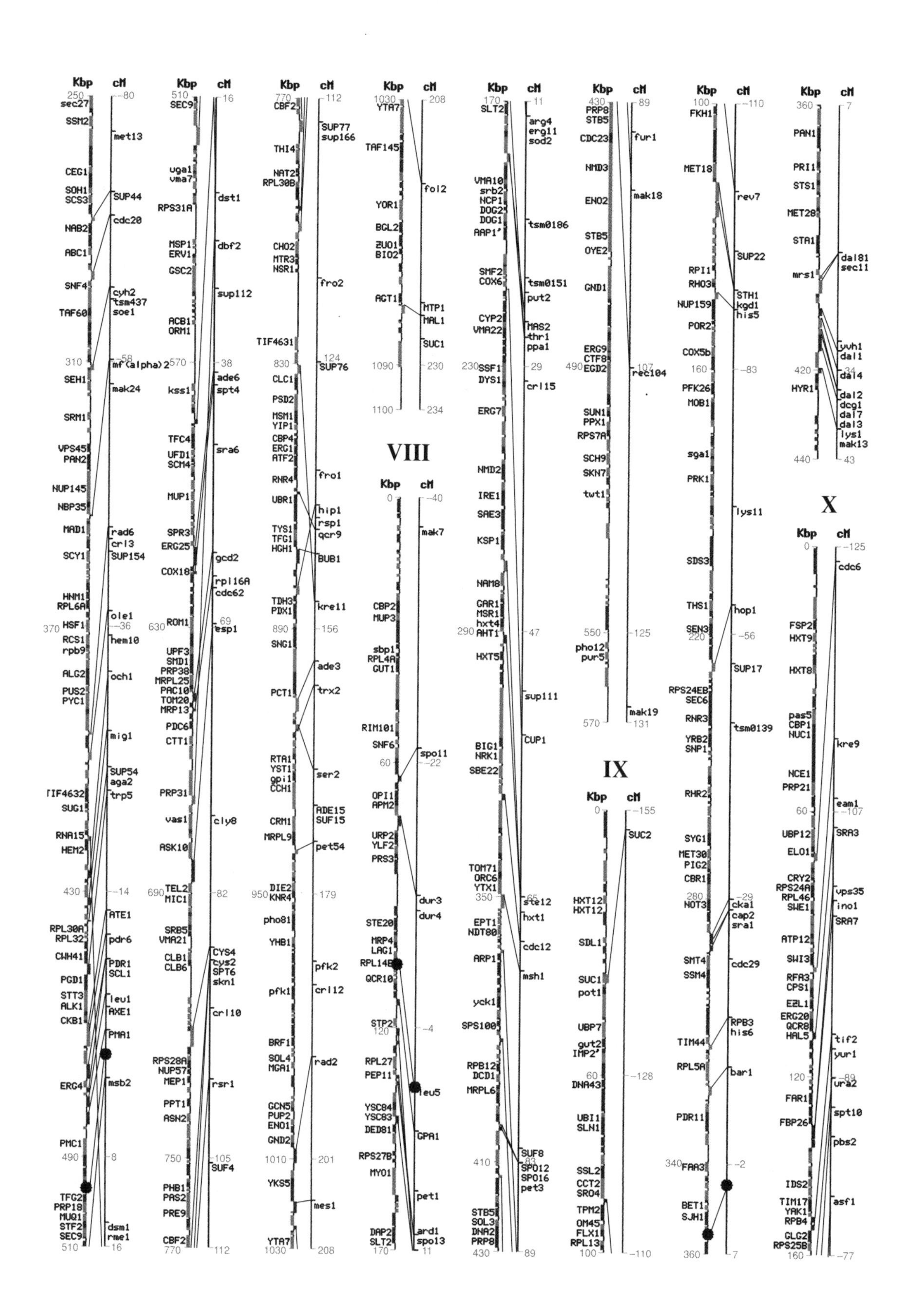
Kbp
cM
VIII
IX
X

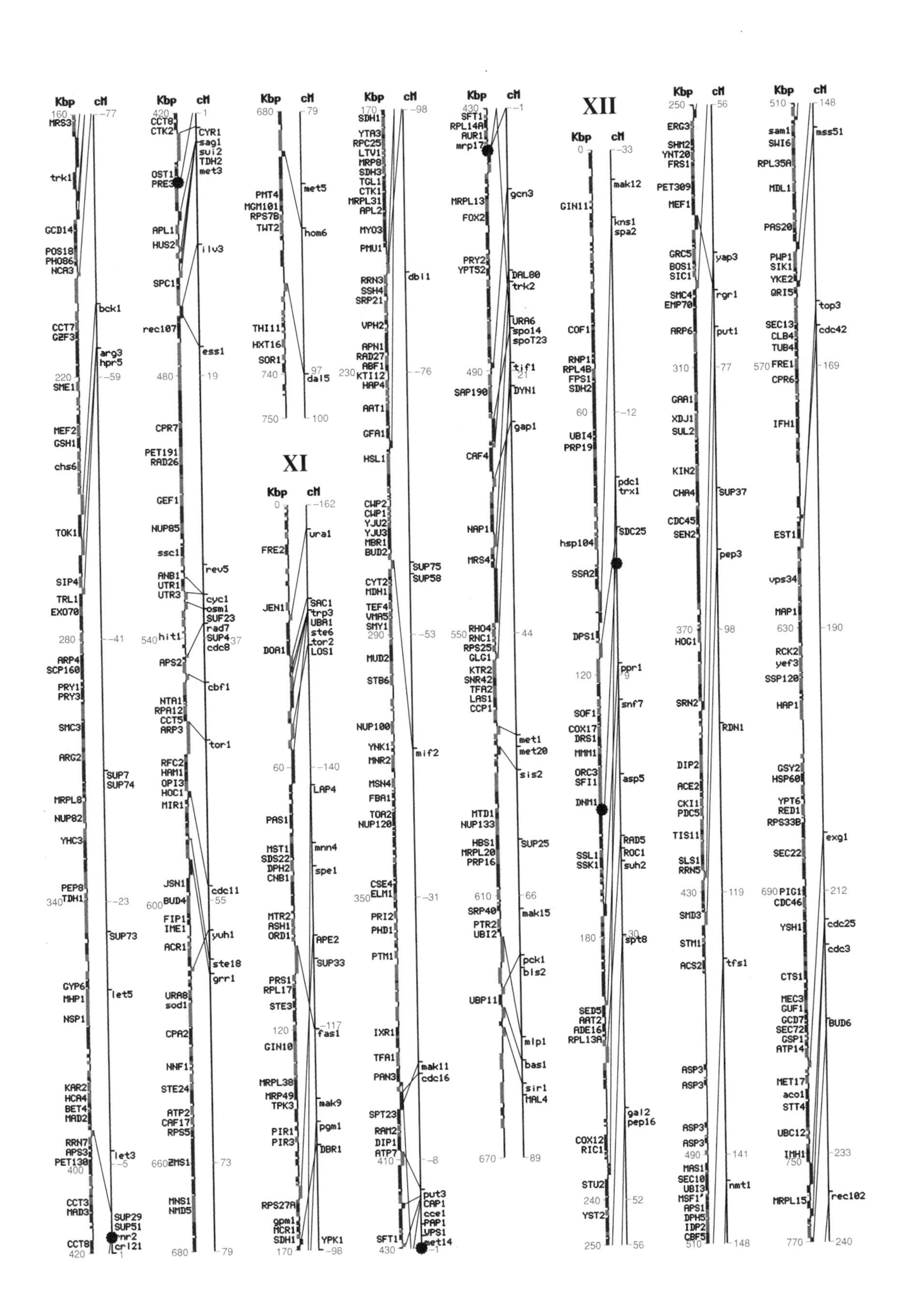

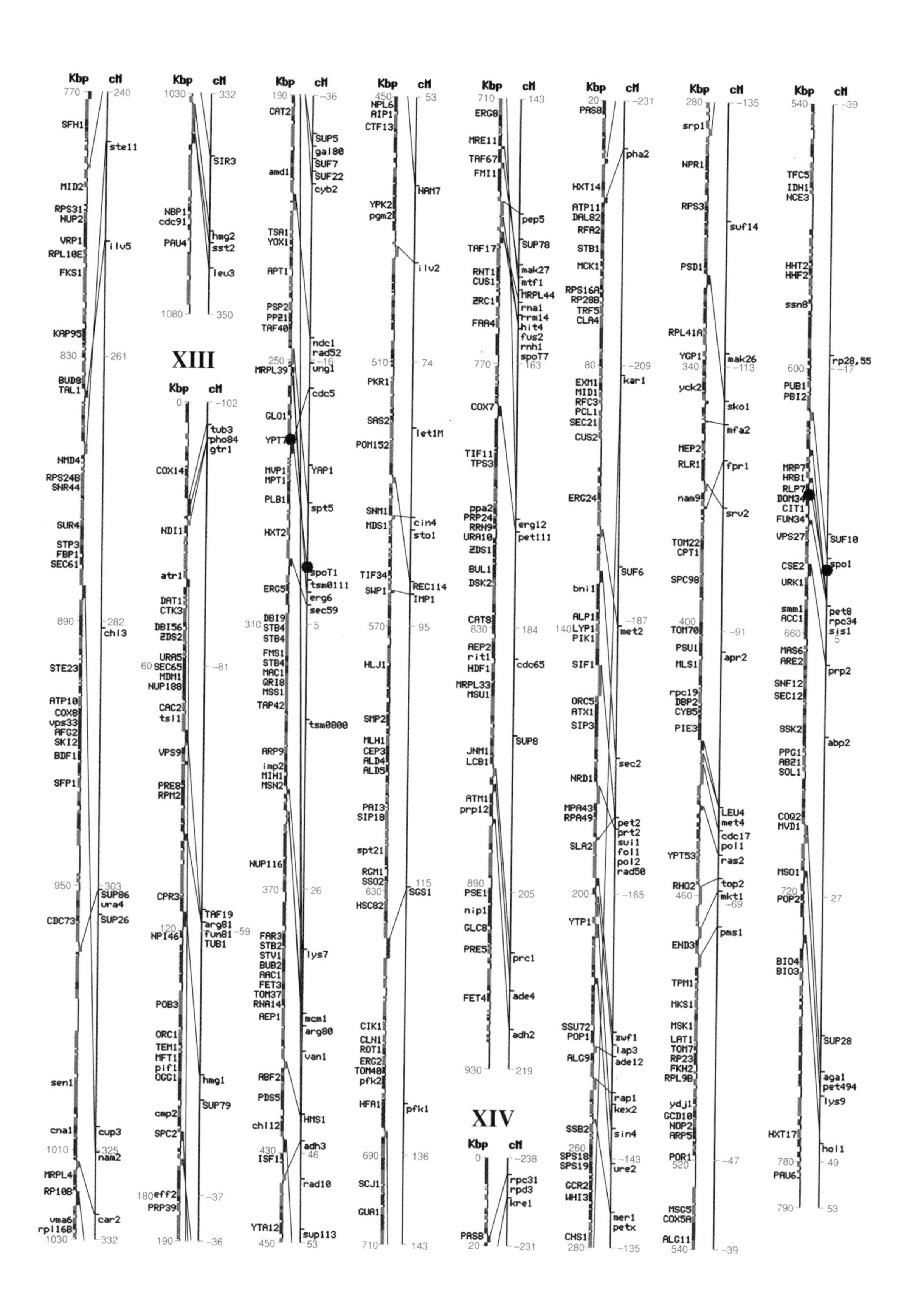
Kbp
cM
770
240
SFH1
ste11
MID2
RPS31
NUP2
VRP1
ilv5
RPL10E
FKS1
KAP95
830
261
BUD8
TAL1
NMD4
RPS24B
SNR44
SUR4
STP3
FBP1
SEC61
890
282
chl3
STE23
ATP10
COX8
vps33
AFG2
SKI2
BDF1
SFP1
950
303
SUP86
ura4
CDC73
SUP26
sen1
cna1
cup3
1010
325
nam2
MRPL4
RP10B
car2
vma6
rpl16B
1030
332
Kbp
cM
1030
332
SIR3
NBP1
cdc91
PAU4
hmg2
sst2
leu3
1080
350
XIII
Kbp
cM
0
-102
tub3
pho84
gtr1
COX14
NDI1
atr1
DAT1
CTK3
DBI56
ZDS2
URA5
60
SEC65
MDM1
NUP188
-81
CAC2
tsl1
VPS9
PRE8
RPM2
CPR3
TAF19
arg81
fun81
TUB1
120
NPI46
-59
POB3
ORC1
TEM1
MFT1
pif1
OGG1
hmg1
SUP79
cmp2
SPC2
180
eff2
PRP39
-37
190
-36
Kbp
cM
190
-36
CAT2
SUP5
gal80
SUF7
SUF22
amd1
cyb2
TSA1
YOX1
APT1
PSP2
PPZ1
TAF40
ndc1
rad52
250
-16
MRPL39
ung1
cdc5
GLO1
YPT7
YAP1
MVP1
MPT1
PLB1
spt5
HXT2
spoT1
tsm0111
ERG5
erg6
sec59
DBI9
310
STB4
5
STB4
FMS1
STB4
MAC1
QRI8
MSS1
TAP42
tsm0800
ARP9
imp2
MIH1
MSN2
NUP116
370
26
FAR3
STB2
STV1
BUB2
AAC1
FET3
TOM37
RNA14
AEP1
lys7
mcm1
arg80
van1
ABF2
PDS5
HMS1
chl12
430
adh3
46
ISF1
rad10
YTA12
sup113
450
53
Kbp
cM
450
53
NPL6
AIP1
CTF13
NAM7
YPK2
pgm2
ilv2
510
74
PKR1
SAS2
let1M
POM152
SNM1
cin4
MDS1
sto1
TIF34
REC114
SWP1
IMP1
570
95
HLJ1
SMP2
MLH1
CEP3
ALD4
ALD5
PAI3
SIP18
spt21
RGM1
SSO2
630
115
SGS1
HSC82
CIK1
CLN1
ROT1
ERG2
TOM40
pfk2
HFA1
pfk1
690
136
SCJ1
GUA1
710
143
Kbp
cM
710
143
ERG8
MRE11
TAF67
FMI1
pep5
SUP78
TAF17
mak27
RNT1
mtf1
CUS1
MRPL44
ZRC1
rna1
rrm14
FAA4
hit4
fus2
rnh1
spoT7
770
163
COX7
TIF11
TPS3
ppa2
PRP24
erg12
RRN9
pet111
URA10
ZDS1
BUL1
DSK2
CAT8
830
184
AEP2
rit1
HDF1
cdc65
MRPL33
MSU1
SUP8
JNM1
LCB1
ATM1
prp12
890
205
PSE1
nip1
GLC8
PRE5
prc1
FET4
ade4
adh2
930
219
XIV
Kbp
cM
0
-238
rpc31
rpd3
kre1
PAS8
20
-231
Kbp
cM
20
-231
PAS8
pha2
HXT14
ATP11
DAL82
RFA2
STB1
MCK1
RPS16A
RP28B
TRF5
CLA4
80
-209
kar1
EXM1
MID1
RFC3
PCL1
SEC21
CUS2
ERG24
SUF6
bni1
ALP1
-187
140
LYP1
met2
PIK1
SIF1
ORC5
ATX1
SIP3
sec2
NRD1
MPA43
RPA49
pet2
prt2
sui1
fol1
pol2
rad50
SLA2
200
-165
YTP1
SSU72
POP1
zwf1
lap3
ALG9
ade12
rap1
kex2
SSB2
sin4
260
-143
SPS18
SPS19
ure2
GCR2
WHI3
mer1
petx
CHS1
280
-135
Kbp
cM
280
-135
srp1
NPR1
RPS3
suf14
PSD1
RPL41A
YGP1
mak26
340
-113
yck2
sko1
mfa2
MEP2
RLR1
fpr1
nam9
srv2
TOM22
CPT1
SPC98
400
TOM70
-91
PSU1
apr2
MLS1
rpc19
DBP2
CYB5
PIE3
LEU4
met4
cdc17
pol1
YPT53
ras2
RHO2
top2
460
mkt1
-69
pms1
END3
TPM1
MKS1
MSK1
LAT1
TOM7
RP23
FKH2
RPL9B
ydj1
GCD10
NOP2
ARP5
POR1
520
-47
MSC5
COX5A
ALG11
540
-39
Kbp
cM
540
-39
TFC5
IDH1
NCE3
HHT2
HHF2
ssn8
600
rp28,55
-17
PUB1
PBI2
MRP7
HRB1
RLP7
DOM34
CIT1
FUN34
VPS27
SUF10
CSE2
spo1
URK1
smm1
pet8
ACC1
rpc34
660
sis1
5
MAS6
ARE2
prp2
SNF12
SEC12
SSK2
abp2
PPG1
ABZ1
SOL1
COQ2
MVD1
MSO1
720
POP2
27
BIO4
BIO3
SUP28
aga1
pet494
lys9
HXT17
hol1
780
49
PAU6
790
53

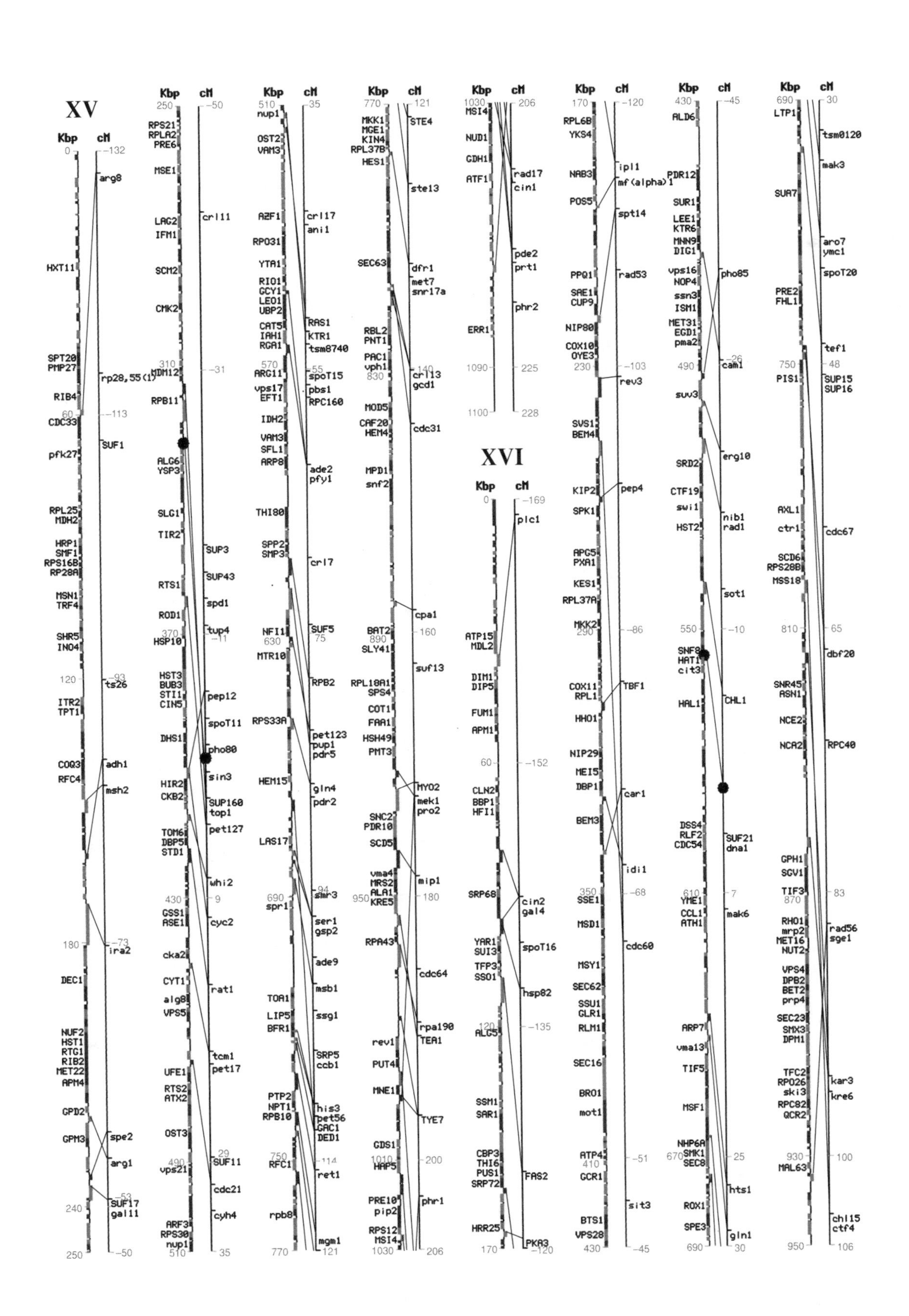
XV
XVI
Kbp
cM

Appendix D

Templates for Making Streak Plates

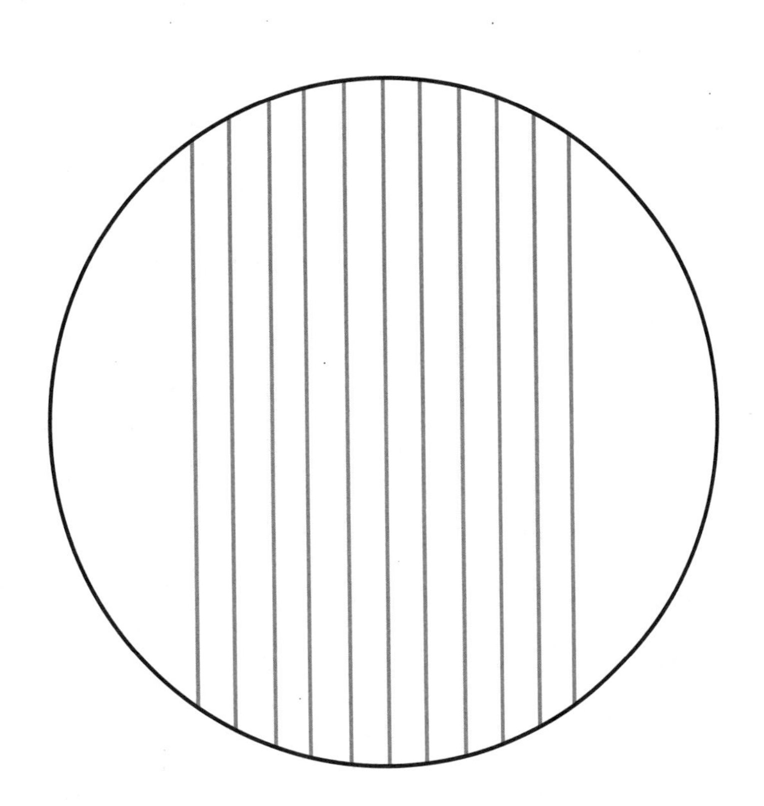

		1	2	3	4		
	5	6	7	8	9	10	
11	12	13	14	15	16	17	18
19	20	21	22	23	24	25	26
27	28	29	30	31	32	33	34
35	36	37	38	39	40	41	42
	43	44	45	46	47	48	
		49	50	51	52		

		1	2	3	4		
	5	6	7	8	9	10	
11	12	13	14	15	16	17	18
19	20	21	22	23	24	25	26
27	28	29	30	31	32	33	34
35	36	37	38	39	40	41	42
	43	44	45	46	47	48	
		49	50	51	52		

		1	2	3	4		
	5	6	7	8	9	10	
11	12	13	14	15	16	17	18
19	20	21	22	23	24	25	26
27	28	29	30	31	32	33	34
35	36	37	38	39	40	41	42
	43	44	45	46	47	48	
		49	50	51	52		

		1	2	3	4		
	5	6	7	8	9	10	
11	12	13	14	15	16	17	18
19	20	21	22	23	24	25	26
27	28	29	30	31	32	33	34
35	36	37	38	39	40	41	42
	43	44	45	46	47	48	
		49	50	51	52		

		1	2	3	4		
	5	6	7	8	9	10	
11	12	13	14	15	16	17	18
19	20	21	22	23	24	25	26
27	28	29	30	31	32	33	34
35	36	37	38	39	40	41	42
	43	44	45	46	47	48	
		49	50	51	52		

		1	2	3	4		
	5	6	7	8	9	10	
11	12	13	14	15	16	17	18
19	20	21	22	23	24	25	26
27	28	29	30	31	32	33	34
35	36	37	38	39	40	41	42
	43	44	45	46	47	48	
		49	50	51	52		

APPENDIX E

Electrophoretic Karyotypes of Strains for Southern Blot Mapping

This appendix has been largely superseded by the sequencing of the yeast genome (see Appendix C, Yeast Genetic and Physical Maps), but this information is still useful for evaluating the size of whole chromosomes by pulsed-field gel electrophoresis.

CHROMOSOME ASSIGNMENTS

Adapted from Carle and Olson (1985).

The following are indexed by band (in order of decreasing electrophoretic mobility).

Band	Chromosome	Specific identifying hybridization probe
1	I	CDC19
2	VI	SUP11
3	III	SUP61
4	IX	SUP17
5A	VIII	ARG4
5B	V	URA3
6	XI	URA1
7	X	URA2
8	XIV	SUF10
9	II	LYS2
10A	XIII	SUP8
10B	XVI	GAL4
11A,B	XV, VII	SUP3, LEU1
12	IV	SUP2

DEFINING STRAINS

Strain	Resolved doublet	Genotype
AB972	(Standard)	*MATα trp1*
A364a	5A, 5B	*MAT***a** *ura1 lys2 ade2 ade1 his7 tyr1*
YPH45	10A, 10B	*MAT***a** *ura3-52 lys2 ade2 trp1-Δ1*

MAPPING STRAINS

Strain	Genotype
YPH80	*MATα ura3-52 lys2 ade⁻ his7 trp1-Δ1*
YPH81	*MAT***a** *ura3-52 lys2 ade⁻ trp1-Δ1*

YPH149 and YPH152 are derived from YPH80 and YPH81, respectively, via two chromosome-fragmentation events. They have the following differences from the standard karyotype:

1. Band 11 equals chromosome XV only (no VII).
2. A chromosome fragment ($URA3^{+}$) derived from chromosome VII and carrying sequences centromere-proximal to RAD2 migrates between bands 10B and 11.
3. A chromosome fragment ($TRP1^{+}$) derived from chromosome VII and carrying sequences centromere-distal to RAD2 migrates below band 1.

Note that YPH149 and YPH152 grown in minimal medium (selecting URA^{+}) show a more intense 90-kb chromosome fragment.

REFERENCE

Carle G.F. and Olson M.V. 1985. An electrophoretic karyotype for yeast. *Proc. Natl. Acad. Sci.* **82:** 3756–3760.

Appendix F

Strains

Experiment I		Looking at Yeast Cells
1-1	FY86	*MATα ura3-52 leu2Δ1 his3Δ200*
1-2	FY23	*MAT***a** *ura3-52 leu2Δ1 trp1Δ63*
1-3	FY23 x 86	*MAT***a***/α ura3-52/ura3-52 leu2Δ1/leu2Δ1 trp1Δ63/TRP1 HIS3/his3Δ200*
1-4	FY23 x 86 x [pTD125]	*MAT***a***/α ura3-52/ura3-52 leu2Δ1/leu2Δ1 trp1Δ63/TRP1 HIS3/his3Δ200* [pTD125]
1-5	FY23 x 86 x [pDAb204]	*MAT***a***/α ura3-52/ura3-52 leu2Δ1/leu2Δ1 trp1Δ63/TRP1 HIS3/his3Δ200* [pDAb204]
1-6	FY23 x 86 x [pTY20]	*MAT***a***/α ura3-52/ura3-52 leu2Δ1/leu2Δ1 trp1Δ63/TRP1 HIS3/his3Δ200* [pTY20]

Experiment II		Isolation and Characterization of Auxotrophic, Temperature-sensitive, and Osmotic-sensitive Mutants
2-1	S288C	*MATα mal gal2*
2-2	D665-1A	*MAT***a**

Experiment III		Meiotic Mapping
3-1	GRY2501	*MAT***a**, *ura3-52, leu2-3-112, arg4-Δ42, trp2, cyh2*
3-2	GRY2502	*MATα, ura3-52, trp1-289, arg4-ΔBgl*II, *ade1*
3-3	3-1 x 3-2	
3-4	AAY1018	*MAT***a**, *his1*
3-5	AAY1017	*MATα, his1*
3-6	GRY2506	*MAT***a**, *arg4-Δ42, his3Δ1, trp1-289*
3-7	GRY2507	*MATα, arg4-Δ42, his3Δ1, trp1-289*
3-8	GRY2508	*MAT***a**, *arg4-ΔBgl*II, *his3Δ1, trp1-289*
3-9	GRY2509	*MATα, arg4-ΔBgl*II, *his3Δ1, trp1-289*
3-10	GRY2510	*MAT***a**, *trp1-289, ura3-52*
3-11	GRY2511	*MATα, trp1-289, ura3-52*
3-12	GRY2512	*MAT***a**, *trp2, ura3-52*
3-13	GRY2513	*MATα, trp2, ura3-52*

Experiment IV		Mitotic Recombination and Random Spore Analysis
4-1	TSY812	*MATα can1 hom3 leu2 lys2 ura3*
4-2	TSY813	*MAT***a** *ade2 his1 lys2 trp1*
4-3	GRY2426	*MAT***a** *his3Δ1 met15Δ0*
4-4	GRY2427	*MATα his3Δ1 met15Δ0*

Experiment V		Transformation of Yeast
5-1	TSY623	MATα ade2-101 his3-Δ200 leu2-3,112 ura3-52
5-3	TSY1017	MATa his3-Δ200 leu2-3,112 trp1-1 ura3-52
5-4	TSY808	MATa lys2-801

Experiment VI		Synthetic Lethal Mutants
6-1	BY4741	MATa ura3Δ0 his3Δ0 leu2Δ0 met15Δ0 orf::G418 (deletion collection)
6-2	2466-1	MATα ura3Δ0 his3Δ0 leu2Δ0 lys2Δ0 cyh2 can1::P_{STE2}-$his5^+$ mad2::NAT

Experiment VII		Gene Replacement
7-1	BY4741	MATa his3Δ1 leu2Δ0 ura3Δ0 met15Δ0
7-2	2404	MATα ade2 trp1 leu2 ura3 his3 lys2-801 ndc10::TRP1 [pRG68]
7-3	BY4741	MATa his3Δ1 leu2Δ0 ura3Δ0 met15Δ0

Experiment VIII		Isolation of *ras2* Suppressors
8-1	FY87	MATα ura3-52 leu2Δ1 his3Δ200
8-2	DAY229	MATa ura3-52 leu2Δ1 trp1Δ63 ras2Δ0::kan^r
8-3	DAY230	MATα ura3-52 leu2Δ1 his3Δ200 ras2Δ0::kan^r

Experiment IX		Manipulating Cell Types
9-1	NE2	MATα ura3-52, leu2-3,112
9-2	AAY1017	MATα his1
9-3	AAY1018	MATa his1
9-4	YSC006	MATα ura3 ade2-1 trp1-1 can1-100 leu2-3,112 his3-11,15 $[psi^+]GAL^+$
9-5	YSC005	MATa ura3 ade2-1 trp1-1 can1-100 leu2-3,112 his3-11,15 $[psi^+]GAL^+$

Experiment X		Isolating Mutants by Insertional Mutagenesis
10-1	BY4741	MATa his3Δ1 leu2Δ0 ura3Δ0 met15Δ0

Experiment XI		Isolation of Separation of Function Mutants by Two-hybrid Differential Interaction Screening
11-1	Y187	MATα gal4 gal80 his3 trp1-901 ade2-101 ura3-52 leu2-3,112 cyh^r P_{GAL}-lacZ
11-2	Y190	MATa gal4 gal80 his3 trp1-901 ade2-101 ura3-52 leu2-3,112 URA3:: P_{GAL}-lacZ LYS2:: P_{GAL}-HIS3 cyh^r
11-3	Y187	MATα gal4 gal80 his3 trp1-901 ade2-101 ura3-52 leu2-3,112 cyh^r P_{GAL}-lacZ [pAIP70]
11-4	Y187	MATα gal4 gal80 his3 trp1-901 ade2-101 ura3-52 leu2-3,112 cyh^r P_{GAL}-lacZ [pJT20]

Appendix G

Counting Yeast Cells with a Standard Hemocytometer Chamber

(A. Kistler and S. Michaelis)

Using a 10x microscope objective (and a 10x ocular), the "large" square shown in the circle below will fill your field of view. This square traps a volume of 0.1 μl. Therefore, to calculate the number of cells/ml in a particular culture, use the following formula:

cells/square x 10^4 x dilution factor = # cells/ml in your culture

STANDARD HEMOCYTOMETER CHAMBER

Modified, with permission, from Sigma-Aldrich Co. (©1994).

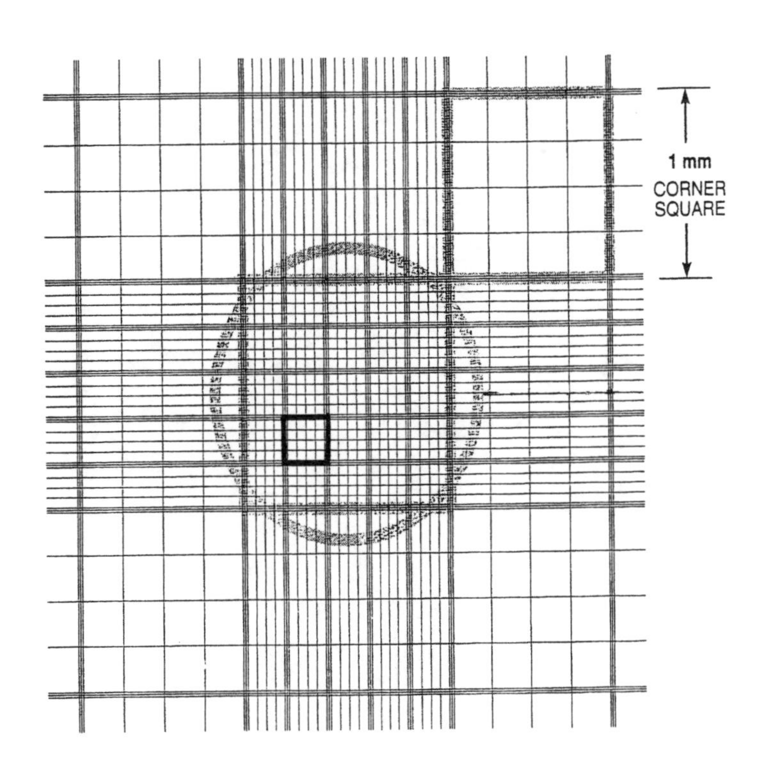

A general guide to typical cell counts for yeast cultures:

Culture	OD_{600}	Cell count	Dilution to count with hemocytometer
Sat'd YEPD	25	3 x 10^8/ml	10^{-2} or 10^{-3}
Log YPD	0.25	3 x 10^6/ml	10^0 or 10^{-1}
Sat'd SC-LYS	5	1 x 10^8/ml	10^{-1} or 10^{-2}

Generally, it is optimal to use a 40x microscope objective, in which case only a portion of the circled field is visible, i.e., several "small" squares (a small square is indicated in bold at lower left in diagram on p. 227). Then, count the cells in five small squares and use the following formula:

cells in 5 "small" squares x 5 x 104 x dilution factor = # cells/ml in your culture

APPENDIX H

Tetrad Scoring Sheet

(see next page)

Strain/Genotype									
A									
B									
C									
D									
A									
B									
C									
D									
A									
B									
C									
D									
A									
B									
C									
D									
A									
B									
C									
D									
A									
B									
C									
D									
A									
B									
C									
D									
A									
B									
C									
D									
A									
B									
C									
D									
A									
B									
C									
D									
A									
B									
C									
D									